Progress in Enzyme and Ion-
Selective Electrodes, Lubbers,
Acker, Buck, Eisenman, Kessler,
Simon, eds. 1981

# Progress in Enzyme and Ion-Selective Electrodes

Edited by D. W. Lübbers, H. Acker
R. P. Buck, G. Eisenman, M. Kessler
and W. Simon

With 98 Figures

Springer-Verlag
Berlin Heidelberg New York 1981

Prof. Dr. Dietrich Werner Lübbers
Priv. Doz. Dr. Helmut Acker
Max-Planck-Institut für Systemphysiologie, Rheinlanddamm 201
4600 Dortmund

Prof. Dr. Richard Pierson Buck
University of North Carolina, Department of Chemistry
Chapel Hill, NC 27514/USA

Prof. Dr. George Eisenman
University of California, Department of Physiology, The Center
for the Health Sciences, Los Angeles, CA 90024/USA

Prof. Dr. Manfred Dietrich Kessler
Universität Erlangen, Institut für Physiologie und Kardiologie,
Waldstraße 6, 8500 Erlangen

Prof. Dr. Wilhelm Simon
Institute of Technology, Department of Organic Chemistry,
Universitätsstraße 16, CH-8092 Zürich

Proceedings of the Meeting on Theory and Application of
Ion-Selective Electrodes in Physiology and Medicine, held at
Dortmund on July 28–30, 1980

ISBN 3-540-10499-2 Springer-Verlag Berlin Heidelberg New York
ISBN 0-387-10499-2 Springer-Verlag New York Heidelberg Berlin

Library of Congress Cataloging in Publication Data. Main entry under title:
Progress in enzyme and ion-selective electrodes. "Proceedings of the
meeting on Theory and Application of Ion-Selective Electrodes in Physiology
and Medicine, held at Dortmund on July 28–30, 1980." 1. Electrodes, Ion
selective–Congresses. 2. Electrodes, Enzyme–Congresses. 3. Biological
chemistry–Technique–Congresses. I. Lübbers, D. W. [DNLM: 1. Biomedical
engineering–Congresses. 2. Electrochemistry–Congresses. 3. Electrodes–
Congresses. 4. Enzyme tests–Congresses. 5. Ion exchange–Congresses.
QT 34 P9635 1980]. QP519.9.E43P76. 599.01'925'028. 80-27595



Printing and bookbinding: Beltz, Offsetdruck, Hemsbach
2127/3140-543210

# Foreword

The sciences employing and the technologies producing ion-selective electrodes are in a period of rapid expansion and development. It has been the
purpose of this meeting to reflect upon progress made so far, and through
the diverse contributions and extensive discussions of the participants to
appreciate, the extend of so many different lines of interrelated research;
and by drawing together such a variety of view points, to get a clearer
and more unified understanding of the directions in which further progress
will be likely to bring us.

Such aims are in part limited by the relatively small number of participants essential for an individual involvement by each member, but we think
that we were able to attract a representative group. Since the basic phenomena of our topic are discussed thoroughly in several monographs, the new
results are of value only if their publication takes place almost immediately. Hereby necessary sacrifices in the unity of the use of symbols and in
the volume of published discussions must be tolerated.

The significant advances made between this and the two previous ion-selec-
tive electrode meetings held in 1974 at Schloß Reisenburg (Ion and Enzyme
Electrodes in Biology and Medicine (1976). Eds. M. Kessler, L.C. Clark Jr.,
D.W. Lübbers, I.A. Silver, W. Simon. Urban & Schwarzenberg, München-Berlin-
Wien) and in 1977 at the Max-Planck-Institut at Dortmund (Theory and Application of Ion-Selective Electrodes in Physiology and Medicine (1978).
Eds. D.W. Lübbers, G. Eisenman, M. Kessler, W. Simon. Arzneim-Forsch (Drug
Res) 28: 705-717 and 866-883) give cause for anticipation of further in-
novations. In order to continue to provide a forum for participation in
debate, the sharing of discoveries and ideas along with continuing to chart
the progress of the field, we hope to be able to convene another meeting
at some further data.

We are greatly indebted to the Deutsche Forschungsgemeinschaft and the
Max-Planck-Gesellschaft zur Förderung der Wissenschaften who provided the
funds enabling the realization of the meeting. We also wish to thank
Hoffmann-La Roche & Co. Ltd. for additional generous financial support.

On the behalf of the Organizing Committee, H. Acker, R.P. Buck, G. Eisenman,
M. Kessler and W. Simon, I would like to thank Mrs. G. Blümel and the other
members of the Max-Planck-Institut für Systemphysiologie who assisted in the
preparation, organization and conduction of the meeting and in the edition
of this book. I would also like to acknowledge the great help in editing
which was given to us by Mr. A.J. Baker.


D.W. Lübbers

# Table of Contents

# Participants

Priv. Doz. Dr. H. Acker, Max-Planck-Institut für Systemphysiologie, Rhein-
landdamm 201, D-4600 Dortmund / FRG
Dr. H. R. Ahmad, Ruhr-Universität Bochum, Institut für Physiologie, Lehr-
stuhl I, Postfach 2148, D-4630 Bochum-Querenburg / FRG
Dr. D. Ammann, Swiss Federal Institute of Technology, Dept. of Organic Che-
mistry, Universitätsstr. 16, CH-8092 Zürich, Switzerland
Prof. Dr. W. McD Armstrong, Indiana University School of Medicine, 1100
West Michigan St., Indianapolis, Indiana 46223 / USA
. Attoe, Corning Medical, Colchester Road, Halstead, Essex  England
A. Baker, University of Bristol, Dept. of Physiology, Medical School, Uni-
versity Walk, Bristol BS8 1TD, England
Dr. K. Ballanyi, Ruhruniversität Bochum, Lehrstuhl für Tierphysiologie,
D-4630 Bochum / FR
Dr. N. Bindslev, University of Copenhagen, Institute of Medical Physio-
logy, Panum Institute, Dk-2200 Copenhagen N / Denmark
Priv. Doz. Dr. Dr. U. Borchard, Universität Düsseldorf, Pharmakologisches
Institut, Moorenstrasse 5, D-4000 Düsseldorf / FRG
Prof. Dr. G. ten Bruggencate, Universität München, Physiologisches Insti-
tut, Pettenkoferstr. 12, D-8000 München 2 / FRG
Prof. Dr. R. P. Buck, University of North Carolina, Dept. of Chemistry,
Chapel Hill, N.C., 27514 / USA
B.Sc. T. Byrne, Rhein.-Westf. Technische Hochschule Klinikum Aachen, Goe-
thestr. 27-29, D-5100 Aachen / FRG
Prof. Dr. K. Cammann, Universität Ulm, Abt. f. Analytische Chemie, Oberer
Eselsberg, D-7900 Ulm / FRG
Dr. J. W. Deitmer, Ruhr-Universität Bochum, Abt. Biologie, D-4630 Bochum
/ FRG
Dr. M. Delpiano, Max-Planck-Institut für Systemphysiologie, Rheinlanddamm
201, D-4600 Dortmund / FRG
Dr. H.M. Einwächter, Universität Bonn, Physiologisches Institut II, Wil-
helmstr. 31, D-5300 Bonn 1 / FRG
Prof. Dr. G. Eisenman, University of California, Dept. of Physiology, The
Center for the Health Sciences, Los Angeles, Ca.90024 / USA
Dr. D. Ellis, Dept. of Physiology, University Medical School, Teviot Place,
Edinburgh EH 8 9AG / Scotland
Dr. M. Fischer, Max-Planck-Institut für Systemphysiologie, Rheinlanddamm
201, D-4600 Dortmund 1 / FRG
Prof. Dr. E. Frömter, Max-Planck-Institut für Biophysik, Kennedyallee 70
D-6000 Frankfurt/Main / FRG
Dr.W. R. Galey, University ot New Mexico School of Medicine, Dept. of
Physiology, Albuquerque, New Mexico 87131 / USA
Dr. L. S. Gettes, University of North Carolina, Division of Cardiology
349 Clinical Sciences Bldg. 229H, Chapel Hill, N.C 27514
Dr. M. Güggi, Universität Erlangen-Nürnberg, Institut für Physiologie und
Kardiologie, Waldstr. 6, D-8520 Erlangen / FRG
Prof. Dr. F. J. Haberich, Institut für Angewandte Physiologie der Philipps-
Universität Marburg, Lahnberge, D-3550 Marburg/Lahn /FRG

Dr. A. Jon Hansen, University of Copenhagen, Institute of Medical Physiology Dept. A, The Panum Institute, Blegdamsvej 3C, Dk-2200 Copenhagen N / Denmark

Dr. D. K. Harrison, The University, Dundee DD1 4HN/Scotland

Dr. N. C. Hebert, Microelectrodes Inc., Grenier Industrial Village Londonderry NH 03053 /USA

Prof. Dr. Hj. Hirche, Universität Köln, Lehrstuhl f. Angewandte Physiologie, Robert-Koch-Str. 39, D-5000 Köln 41 /FRG

Dr. G. Hofmeier, Max-Planck-Institut für Psychiatrie, Abt. Neurophysiologie, Kraepelinstr. 2, D-8000 München 40 / FRG

Dr. J. Janata, University of Utah, Dept. of Bioengineering, Salt Lake City, Utah 84112 / USA

Dr. M. Kallerhoff, Universität Göttingen, Physiologisches Institut I, Humboldtallee 7, D-3400 Göttingen / FRG

Prof. Dr. R. Kaufmann, Universität Düsseldorf, Physiologisches Institut, Lehrstuhl f. klinische Physiologie, Moorenstrasse 5, D-4000 Düsseldorf / FRG

Dr. U. Kebbel, Universität Köln, Institut für Angewandte Physiologie, Robert-Koch-Str. 39, D-5000 Köln 41 / FRG

Prof. Dr. M. Kessler, Universität Erlangen, Institut für Physiologie und Kardiologie, Weststr. 6, D-8500 Erlangen / FRG

Prof. Dr. R. N. Khuri, American University, Faculty of Medicine, Beirut / Lebanon

Dr. M. Korth, Technische Universität München, Institut für Pharmakologie und Toxikologie, Biedersteiner Str. 29, 8000 München/FRG

Dr. J. Küppers, Westf. Wilhelms-Universität Münster, Lehrstuhl für Neurophysiologie, Zoologisches Institut, Hüfferstr. 1, D-4400 Münster / FRG

Priv. Doz. Dr. E. Leniger-Follert, Max-Planck-Institut für Systemphysiologie, Rheinlanddamm 201, D-4600 Dortmund / FRG

Dr. R. Linton, Dept. of Anaesthetics, St. Thomas's Hospital London SE1 1EH / England

Dr. J. Lohse, Städt. Krankenhaus München-Schwabing, 5. Med. Abt., Kölner Platz 1, D-8000 München 40 / FRG

Prof. Dr. D.W. Lübbers, Max-Planck-Institut für Systemphysiologie , Rheinlanddamm 201, D-4600 Dortmund / FRG

Dr. A. Luttmann, Ruhruniversität Bochum, Physiologisches Institut, Lehrstuhl I, Postfach 2148, D-4630 Bochum / FRG

Dr. J. Machek, Institute of Physiology, Videnská  1083, 14220 Prague 4 / CSSR

Dr. E. Marcoll, Dräger Werk AG, Moislinger Allee 53-55, D-2400 Lübeck

Prof. Dr. W. Meesmann, Universitätsklinikum der Gesamthochschule Essen, Institut für Pathophysiologie, Hufelandstr. 55, D-4300 Essen / FRG

Dr. R. Meier, Universität Bonn, Institut für Cytologie und Mikromorphologie, Ulrich-Haberland-Str. 61a, D-5300 Bonn / FRG

Dr. W. Morf, Swiss Federal Institute of Technology, Dept. of Organic Chemistry, Universitätsstrasse 16, CH-8092 Zürich / Switzerland

Dr. P. Mückenhoff, Ruhr-Universität Bochum, Institut für Physiologie, Lehrstuhl I, Postfach 2148, D-4630 Bochum-Querenburg / FRG

Dr. Ch. Nicholson, New York University Med. Centre, Dept. of Physiology and Biophysics, 550 First Ave., New York, N.Y. 100116/USA

Dr. N. Opitz, Max-Planck-Institut für Systemphysiologie, Rheinlanddamm 201, D-4600 Dortmund / FRG

Dr. H. F. Osswald, F. Hoffmann-La Roche & Co. Ltd. Bioelektronisches Dept., Grenzacher Str. 124, CH-4002 Basel / Switzerland

J. M. Otten, Cordis Europa NV, Roden / The Netherlands

Prof. Dr. E. Rosenberg, Dept. of Physiology, College of Medicine, Howard University, Washington DC, 20059 / USA

Dr. H. Rosskopf, Braun Melsungen AG, Abt. EM, Karl-Braun-Str. 1, D-3508 Melsungen / FRG

Dr. P. Schlieper, Universität Düsseldorf, Pharmakologisches Institut, Moorenstr. 5, D-4000 Düsseldorf /FRG

Prof. Dr. W. Simon, Institute of Technology, Dept. of Organic Chemistry, Universitätsstr. 16, CH-8092 Zürich / Switzerland

H.J. Springer, Universität Düsseldorf, Institut für Pharmakologie, Universitätsstr. 1, D-4000 Düsseldorf / FRG

Dipl. Phys. S. Stiller, Rhein.-Westf. Technische Hochschule, Klinikum Aachen, Goethestr. 27-29, D-5100 Aachen / FRG

Prof. Dr. W. Stockem, Universität Bonn, Institut für Cytologie und Mikromorphologie, Ulrich-Haberland-Str.61a, D-5300 Bonn 1 / FRG

Dr. P. Streichhan, Max-Planck-Institut für Systemphysiologie, Rheinlanddamm 201, D-4600 Dortmund 1 / FRG

Dr. T. Treasure, Brompton Hospital, Fulham Road, London SW 3

Dr. M. Ullrich, Universität München, Physiologisches Institut, Pettenkoferstr. 12, D-8000 München 2 / FRG

Dr. R. D. Vaughan-Jones, University of Oxford, Dept. of Physiology South Parks Road, Oxford OX1 3QT / England

Prof. Dr. J. L Walker, University of Utah College of Medicine, 410 Chipeta Way, Research Park, Salt Lake City, Utah 84108/USA

Dr. H.G. Wolpers, Universität Göttingen, Physiologisches Institut, Humboldtallee 7, D-3400 Göttingen /FRG

Dr. H. R. Wuhrmann, F. Hoffmann-La Roche & Co. Ltd, Bioelectronics Dept., Grenzacher Str. 124, CH-4002 Basel / Switzerland

# The Ability of the Peptide Backbone to Bind Anions as well as Cations – Implications for Peptide Carriers, Channels and Electrodes

G. EISENMAN, R. MARGALIT, K.-H. KUO

The cation-complexing properties of carbonyl and ether oxygen ligands are well known (7, 6, 15, 18) and have led to bulk electrodes for $Li^+$ and $Na^+$ (1, 7) as well as to carrier mediated permeation of theoretically interesting lipid bilayer membranes (2, 10, 11). Less well known is the ability of such groups also to form complexes with anions (4, 11). As a prototype for specific anion-as well as cation-binding to the peptide backbone we present some results of recent studies on Simon's neutral $Li^+$ and $Na^+$ selective complexones (1, 7) whose only ligands are two tertiary amides and two ether groups.

## The $Na^+$-Selective Complexone

The Na-selective complexone N,N,N',N',-Tetrabenzyl-3,6-dioxaoctan-diamide (2) is a non-cyclic neutral molecule with 2 amide carbonyls and 2 ether ligands and produces $Na^+$-selective permeation of GMO bi-layers (2, 10). The membrane conductance observed in the presence of this ligand is proportional to the 1st and 2nd powers of cation and ligand concentration, respectively, which suggests that 2 to 1 ligand to cation complexes are the permeant species. The typical 'hyperbolic' conductance-voltage characteristics and the voltage independent permeability ratios suggest equilibrium interfacial reactions for this molecule with all the ionic species studied so far, consistent with the flexibility expected for 2:1 complexes. Induced permeation of small anions ($Cl^-$ and $NO_3^-$) has been observed with the $Na^+$ complexone which gives rise to an apparent concentration dependence of the permeability ratios; but this complexity disappears when one allows for anion as well as cation permeability in the Goldman Hodgkin Katz equation. The data are then well fitted by permeability ratios which are voltage-independent and are also concentration-independent at least at low ionic concentrations. Limiting values at low concentrations for the permeability ratios are $Na^+(1) > NO_3^-(0.21) > K^+(0.18) > Cl^-(0.103) > Rb^+(0.097) > Cs^+ (0.038) > Li^+ (0.017)$. The selectivity implied by the limiting values of permeability ratios for cations in bilayers is the same as that observed for membrane conductances for cations measured in single salts at high concentrations. This molecule also induces a permeability to polyatomic cations with 2:1 stoichiometry and with permeability ratios being: $Na^+ (1.0) \cong CH_3(CH_2)NH_3^+ (0.98) > C_2H_5NH_3^+ (0.28) > HONH_3^+ (0.16) > NH_4^+(0.13) > CH_3NH_3^+(0.07) > H_2NNH_3 (0.04) > (CH_3)_3NH^+ (0.01) > (CH_3)_4N^+ (0.002)$. The low selectivity for the three-dimensionally bulky $(CH_3)_3NH^+$ and $(CH_3)_4N^+$ suggests an upper limit to the cavity size of the 2:1 complexes. All of the above results, as well as comparison of the conductances in GMO/ decane vs. PE/decane bilayers, indicate that this molecule acts as a classical carrier, forming 2:1 carrier: cation complexes and probably also carrying anions as 2:1 carrier: anion complexes.

2

## The Li$^+$-Selective Complexone

The Li$^+$ selective complexone N,N'-deheptyl-N,N'5"-tetramethyl-3-7-dioxan-onan-diamide, which produces Li$^+$ selective bulk electrodes (7) also acts as an Li$^+$ selective carrier in bilayers (12). Respectable conductances are produced at reasonable concentrations ($2.5 \times 10^{-5} \Omega^{-1} cm^{-2}$ in 1 N $Li_2SO_4$ at an aqueous ligand concentration of $10^{-6}$M); and an ideal Nernst slope is seen for pure Li$^+$ solutions of all anions. This ligand does not carry divalent ions of either sign, which enabled us to study separately purely cation and purely anion permeation, even for relatively poorly permeant species. In mixtures of cations as well as in cation-anion mixtures the membrane potential obeys the Goldman-Hodgkin-Katz equation with concentration independent (and voltage independent) permeability ratios over a rather wide concentration range. The permeability ratio selectivity is: Li$^+$ (1)$>$Tl$^+$(0.25)$>$Na$^+$(0.12)$>$ NH$_4^+$ (0.056)$>$NO$_3^-$(0.047)$>$K$^+$(0.046)$>$Rb$^+$(0.032) $>$Cl$^-$ (0.018)$>$Cs$^+$ 0.016). The same selectivity sequence, with very similar magnitudes, is observed for the ratios of zero current conductances of single salts and is the same as the sequence for those ions studied in bulk electrodes (7).

Using permeant divalent cations to study the anions and using permeant divalent anions to measure the cations, measurements in bilayers enable the complicated stoichiometry of this molecule to be analyzed. Examination of the concentration conductance behavior indicates that this molecule can carry cations as 1:1 or 2:1 ligand-cation complexes. Measurements of dilution potentials verify that the charge of the complex is +1. In contrast this molecule when carrying anions in bilayers, can form 1:1 or 2:2 anion :ligand complexes, whose charge, measured by dilution potentials, is either −1 or −2, respectively. Thus, this molecule not only can act as a primary carrier for anions across bilayer membranes as was found for a related Na$^+$ selective carrier but is the first example of a double negatively charged entity within a lipid bilayer.

Besides its expected Li$^+$ selectivity, an even higher ("supra-Ia" (3)) Ag$^+$ selelectivity is also observed for the Li$^+$ complexone (and, interestingly, also for the neutral channel-forming polypeptide Gramicidin A (13)) which contrasts to the low ("sub Ia" (8)) Ag$^+$ selectivity observed with the ether- (and ester-)containing macrotetrolide actin carriers (3). This suggests that the imide carbonyls are important cationic ligands in the molecules, an inference which is supported by the high sensitivity of the cation selectivity to the nature of the imide substituents. For example, replacing the aliphatic N-imide substituents of the Li$^+$ carrier by benzyl residues in the Na$^+$ carrier results in a selectivity shift among group Ia cations from Li Na K Rb Cs (selectivity sequence XI) to Na$>$K $>$Rb$>$Ca$>$Li (sequence VII). Correspondingly, as the "negative field strength" of the carbonyl decreases, the relative anion to cation selectivity of the Na$^+$ carrier becomes enhanced in comparison with that of the Li$^+$ carrier.

## Anion Binding

The present situation with regard to the possibility of specific anion binding to the uncharged fully H-bonded peptide backbone appears to be similar to the situation a dozen years ago when it was initially surprising to find that the uncharged polypeptide backbone could bind cations without the need for fixed charges. Although, Schleich and Von Hippel (16)

proposed a direct anion binding to the polypeptide backbone at the imide
nitrogen to explain the specific ion effects on the solution conformation
of poly-l-proline, this idea seems not have been pursued further, perhaps
because of a paucity of unambigous experimental data on clear anion effects
on the peptide backbone. Our studies with the present carriers are the
first clear instance of such interactions in carrier systems, but it should
be noted that other signs of specific anion binding to the presumably fully
H-bonded polypeptide backbone of the gramicidin A channel have also recently been reported (5).

The ability of molecules usually thought of as cation carriers to carry
anions under appropriate conditions might be of some interest in the development of practical electrodes. It is certainly of interest from the
purely theoretical point of view. Presumably, the reason that anion selectivity has not been seen in bulk electrodes is that the solvents used
for such electrodes so far have disfavoured anions. Perhaps through the use
of solvents less unfavourable to anions or by the addition of positively
charged lipophilic dopants (e.g. tetraphenyl phosphonium) one might be able
to utilize the anion complexing abilities of ester, amide, and ether containing molecules.


## The Physical Basis for Cation or Anion Binding

It is possible to get some feel for the energies of interactions of cations
and anions with the amide (or imide) nitrogen and the carbonyl groups of
the peptide bond from Somsen's (17) thermochemical data for the enthalpies
of solvation of the alkali halide salts in amide solvents. These energies
for cations and anions, separated using the Halliwell and Nyburg convention, are plotted in Fig. 1 for DMF and formamide, as well as water. DMF
and formamide were chosen as models because the former, like our carrier,
cannot form hydrogen bonds; whereas the amide protons of the latter form
strong hydrogen bonds.

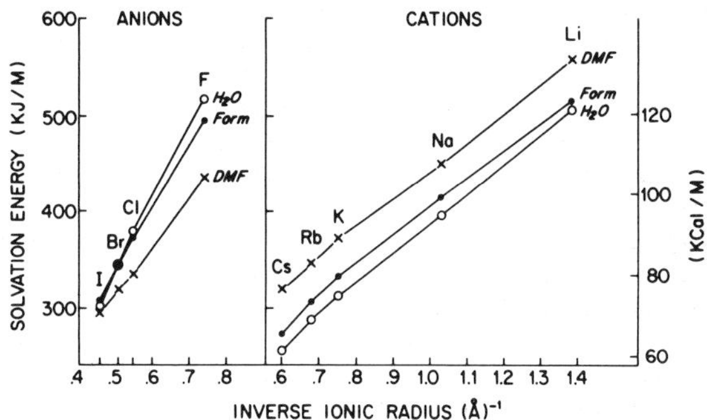


Figure 1. Solvation energies in amide solvents

DMF therefore is an excellent model in which to examine the pure interactions between anions and the tertiary amide nitrogen. These interactions are substantial; indeed DMF solvates (binds) the largest anion,
I⁻, almost as well as water or formamide, which is not surprising since
H-bond energies are generally considered negligible for I⁻. These data
support the Schleich and von Hippel hypothesis of significant tertiary

4

amide nitrogen interaction with anions and indicate that such interactions
can contribute substantially to anion binding interactions in our carrier
molecule.

In contrast if we compare the energies for a strongly H-bonding anion
such as $F^-$, we can assess the magnitude of the extra H-bonding energy
by the differences in solvation energies of $F^-$ in formamide vs. DMF
which is roughly 10 KCal/M (1 Cal= 4.19 Joule). Notice that, on this view,
the extra bonding energy to $F^-$ due to H-bonding in formamide is quite
comparable to that in water as judged by the similar solvation energies
for $F^-$ in formamide and water.

It is possible to examine this in more detail and gain some insight into
the factors underlying cation vs. anion selectivity by comparing the ionic
enthalpies of transfer between the amide solvents and water, as is done in
Fig. 2. Such a procedure is a prototype for the transfer of an ion from
water into the present carrier molecule, as well as for ion interactions
with the polypeptide backbone in aqueous media. From left to right in
Fig. 2, one can compare the selectivity for cations vs. anions for the solvents DMF and formamide, as well as the backbone binding to poly-1-pro-
line II (16) and the binding to the $Li^+$ carrier (4, 11). Ions preferred by
the solvent, backbone, or carrier fall above the dashed horizontal line.
Ions rejected by the solvent, backbone or carrier (i.e. preferred by water)
fall below this line. (Note that in the case of the $Li^+$ carrier it is not
possible to assess an absolute preference but only relative values between
cations and anions). Figure 2 illustrates that a marked preference for cations over anions is expected by a fully H-bonded peptide (or tertiary a-
mide) backbone with no H-bond donating abilities exemplied by DMF. The re-
jection of anions is most extreme in the case of the most strongly hydrogen
bonded species (e.g. $F^-$).

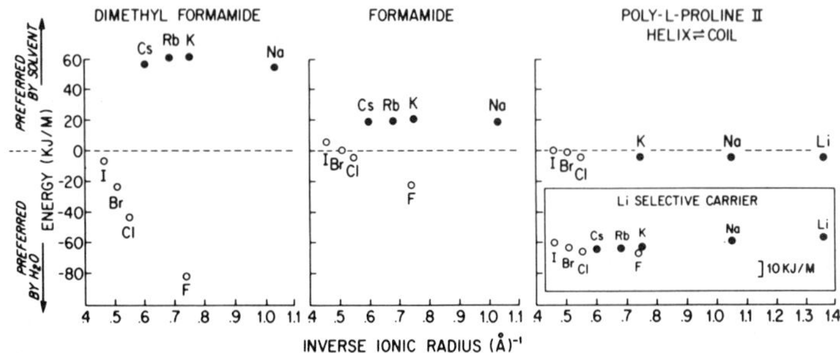


Figure 2.  Selectivity of amide solvents, poly-1-proline, and the $Li^+$ selective carrier


The "Interposed Water Molecule Hypothesis"

The low anion vs. cation selectivity for interactions solely with a tertiary amide group of the non H-bonding solvent DMF, compared with the comparable anion vs. cation selectivity of the $Li^+$ carrier and poly-1-proline at
the right, indicate that the tertiary amide nitrogen of our carrier alone

is unlikely to be the basis for the comparable anion vs. cation selectivity observed for the carrier molecule and poly-1-proline. However, the effect of H-bonding to provide an additional stabilization of anion binding can be seen in Fig. 2 in the increase in anion vs. cation selectivity between DMF and formamide which reflects formamide's ability to form H-bonds with anions. The 60 KJoule/ mole (14 KCal/mole) enhancement of fluoride selectivity in formamide relative to DMF illustrates roughly the magnitude of the additional stabilization energy for anions that can be provided by hydrogen bonding. Since water molecules included in the complexes can supply these H-bonds in the Li carrier and in poly-1-proline, it seems natural to suggest (4) that by interposing a water molecule between the carbonyl oxygen and the anion, we utilize its hydrogen bonding ability, to stabilize the anion complex to amides which cannot supply a proton. At the same time the water screens the stray field of the carbonyl groups that would otherwise be exposed in an anion complex and thus makes the carbonyl unavailable for cation binding.


## Valinomycin Analogous Can Act as Anion Carriers

From the above arguments and the anion-carrying ability observed for Simon's non-cyclic molecules it is natural to predict that cyclic peptide carriers possessing sufficiently large cavities should therefore be able to carry anions. No such anionic permeation has been reported for small carriers like valinomycin, which do not have room in the center for a water molecule together with an anion; but we have tested two cyclic depsipeptides: Hexadecavalinomycin, which is a tetramer of the repeating unit of valinomycin (i.e., cyclo $[DValLLacLValDHylv]_4$) and an imide-ligand-containing analogue cyclo-$[DValLProLValDHylv]_4$. Both these molecules were synthesized by Ivanov's group at the Shemyakin Institute, and they have cavities large enough to include both an ion and a water molecule. Evidence that these molecules can carry anions has been given for perchlorate elsewhere (12). For the imide-containing carrier c$[DValLProLValDHylv]_4$ we observed a first power dependence of zero-current conductance of PE membranes on carrier concentration in the presence of $Mg(ClO_4)_2$. This indicates that a charged monocarrier membrane-permeating complex has been formed. The dilution-potentials for this complex are also found to be purely anionic (12), demonstrating that perchlorate and not magnesium is the transported ion.

It is possible to evaluate separately the effects of cavity size and the effects of imide vs. ester ligands on anionic permeation and selectivity by a comparative study of anionic vs. cationic selectivities of a series of valinomycin analogues. An example is illustrated in Figure 3, where we have compared the ratios of zero-current conductances of percholorate to potassium for valinomycin, hexadecavalinomycin, $[DValLProLValDHylv]_4$ and Simon's $Li^+$ selective carrier (ETH149). Considering first valinomycin and hexadecavalinomycin, identical in their repeating unit and differing only in cavity size, the smaller-sized valinomycin shows extreme preference for $K^+$ over $ClO_4^-$, while the larger-sized hexadecavalinomycin, although still $K^+$-selective, shows a decrease in the extent of the $K^+$ over $ClO_4^-$ perference. Examining next the effects of imides vs. esters, the imide-containing $[DValLProLValDHylv]_4$ shows an increase of the $ClO_4^-$ to $K^+$ preference relative to the ester-containing similar-sized hexadecavalinomycin to the point of crossing over the $ClO_4^-$ being the preferred ion. Notice that the $ClO_4^-$ selectivity over $K^+$ increases considerab-

6

ly for the noncyclic carrier which has both imide groups and an "open cavity".

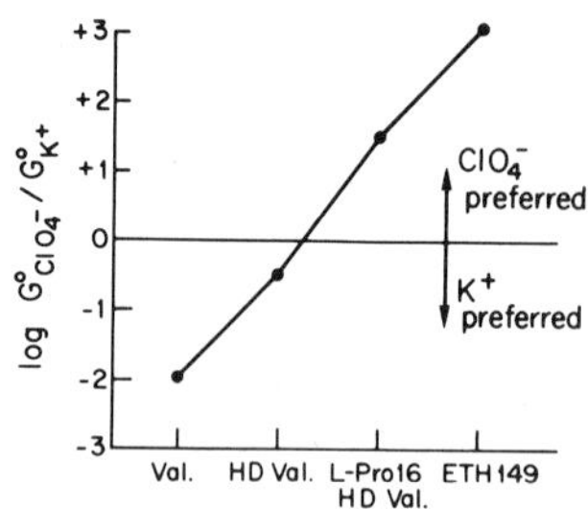



Figure 3. Ratios of zero-current conductances of perchlorate to potassium,
for valinomycin (VAL), hexadecavalinomycin (HDVal), c[DValLProL ValDHlv]$_4$
(L-Pro-16-HdVal) and the Li$^+$-carrier (ETH149); ClO$_4^-$ conductances were measured in PE membranes, K$^+$ conductances were measured in GMO or GDO membranes and "normalizes" to PE membranes, assuming $G^°_{PE}/G^°_{GDO}$ for K$^+$ is 10$^{-3}$
(6)


Implications

We have speculated here in what we will call the "Interposed Water Molecule
Hypothesis" for anion binding to normally cation-preferring ligands that
$H_2O$ molecules, through their ability to form H-bonds simultaneously to the
oxygen ligand and to the anion, provide a method of stabilizing anionic
complexes. This is evidenced not only in Simon's non-cyclic Li$^+$ and Na$^+$
carriers but also in valinomycin analogues, as well as for poly-1-proline
binding. It should be noted that the conformation of the anion complex may
disfavour cations (and vice versa), and it seems likely that cooperative
rearrangements of the backbone and water structure are required in forming
anion vs. cation complexes. This may explain why charged complexes of pure
cations and pure anions have so far been found more easily than mixed symport complexes; and it also has implications for channels whose ion permeation path involves the peptide backbone. For example, when one binds a
cation in the gramicidin channel one must replace a water molecule and must
also set up a particular (and cooperative) orientation of the H-binding
structure of the water (and possibly even of the peptide backbone). Binding
an anion, or its equivalent process of dissociating a proton from a water
to produce an OH$^-$, will set up a different orientation of water molecules.
Clearly, the arrangements of ligands, anions, cations, and water molecules
are independent and only mutually compatible in particular arrangements,
which are likely to be mutually exclusive. This, of course, has implications for the orientations and positions of protons in the extended H-bond
arrays expected in such channels, which suggests that interesting dielectric properties such as ferroelectricity, and piezoelectricity (9,14) might
occur in peptide channels under appropriate conditions of ionic loading.

All present neutral carriers used in Bulk Ion Selective Electrodes for
cations are based on oxygen ligands (peptide, ester, or ether). The ability
of molecules with such ligands to carry anions across the bilayer membrane
is probably relevant to the understanding of "anion interferences" seen in
conventional membrane electrodes such as the valinomycin K$^+$ electrode. We

speculate here that by adjustment of the solvent or through the use of anionic dopants such interferences could be supressed. In addition through the use of cationic dopants the selective anion complexing ability of these molecules might also prove useful for making electrodes specific to certain anions.


<u>References</u>


1. Ammann D, Pretsch E, Simon W (1974) A sodium ion-selective electrode based on neutral carrier. Anal Lett 7: 23
2. Eisenman G (1978) Intervention in: Molecular Movements and chemical reactivity as conditioned by membranes, enzymes and other macromolecules. Adv Chem Phys 39: 316
3. Eisenman G, Krasne S (1975) The ion selectivity of carrier molecules, membranes and enzymes. In: Fox CF (ed) MTP Internat Rev of Sci, Biochem Ser (vol 2), Butterworths, London, p 27
4. Eisenman G, Margalit R (1978) Amphoteric complexes of a neutral ionophore having tertiary amide ligands - a model for anion binding to the polypeptide backbone. In: Leigh JS, Dutton PL, Scarpa A (eds) Frontiers of Biological Energetics (vol 2). Academic Press, New York, p 1215
5. Eisenman E, Sandblom J, Neher E (1978) Interactions in cation permeation through the gramicidin channel Cs, Rb, K, Na, Li, Tl, H, and effects of anion binding. Biophys J 22: 307
6. Eisenman G, Szabo G, Ciani S, McLaughlin SGA, Krasne S (1973) Ion binding and ion transport produced by neutral lipid-soluble molecules. Prog Surf Memb Sci 6: 139
7. Güggi M, Fiedler U, Pretsch E, Simon W (1975) A lithium ion-selective electrode based on a neutral carrier. Anal Lett 8: 857
8. Halliwell HF, Nyburg SC (1963). Enthalpy of hydration of the proton. Trans Faraday Soc 59: 1126
9. Hoshino S, Okaya Y, Pepinsky R (1959) Crystal structure of the ferroelectric phase of $(glycine)_3H_2SO_4$. Phys Rev 115: 323
10. Kuo K-H, Eisenman G (1977) $Na^+$-selective permeation of lipid bilayers mediated by a neutral ionophore. Biophys J 17: 212a
11. Margalit R, Eisenman G (1978) Mode of action of Simon's non-cyclic $Li^+$-selective molecule on bilayers including its ability to carry anions selectively. In: Lübbers DW, Eisenman G, Kessler M, Simon W (eds) Theory and application of ion-selective elctrodes in physiology and medicine. Arzneim Forsch (Drug Res) 28 (I): 707
12. Margalit R, Eisenman G (1979) Some binding properties of the peptide backbone inferred from studies of a neutral non-cyclic carrier having imide ligands. In: Gross E, Meienhofer J (eds) Peptides: Structure and biological function. Pierce Chem Co Publ, New York. p 665
13. McBride D, Szabo G (1978) Blocking of gramicidin channel conductance by $Ag^+$. Biophys J 21: 25a
14. Onsager L (1936) Electric moments of molecules in liquids. J Amer Chem Soc 58: 1486
15. Ovchinnikov YuA, Ivanov VT, Shkrob AM (eds) (1974) Membrane active complexones. Elsevier, Amsterdam.
16. Schleich T, von Hippel P (1969) Specific ion effects on the solution conformation of poly-L-proline. Biopolymers 7: 861
17. Somsen G (1969) Solution and solvation enthalpies of salts in several solvents. Proc 1st Intern Conf Calorimetry and Thermodynamics, Warsaw, p 959

18. Urry DW (1971) Neutral sites for calcium ion binding to elastin and
    collagen: a charge neutralization theory for calcification and its
    relationship to artherosclerosis. Proc Nat Acad Sci USA 68: 810

## Discussion

Buck: In our paper (Buck and Boles) we showed slope changes and peak responses followed by anion responses (negative slope regions) in $K^+$-Valinomycin electrodes. Using diphenylether as the solvent, we subsequently showed (unpublished results) that the peak response (as a measure of anion interference) followed the lipophilicity (Hofmeister series). Solutions of $K^+$, tetraphenylborate were extreme in that only anion response were observed. Pure $K^+$ response was limited to salts of $K^+$ with $F^-$, $OH^-$, divalent anions such as $SO_4^{2-}$. Even KCl showed a reduced slope at 0.1 M Kcl and above.

These results led to the still unresolved problem of $K^+$(val) vs. anion transport in thick membranes. How can electro-neutrality in the thick membranes be maintained, and yet have essentially pure permselectivity for $K^+$ as inferred by the Nernstian response to $K^+$? For example, if in KCl bathing solutions, how can the membrane contain $K^+$val $Cl^-$ and still show good a $K^+$ response without evidence for $Cl^-$ transport? One would expect reduced slopes (at all bathing concentrations) as the transferrence number of $K^+$ is lowered by some non-zero transport number for $Cl^-$.

The "peaking" response correlation with lipophilicity, presumably has nothing to do with anion bonding through nitrogen, or intervention of $H_2O$. The Hofmeister series is thought to be a low-field polarizability effect which applies to anion solubilities in many low-dielectric organic solvents. It is a question whether this $H_2O$ intervention may be involved in the interpretation of lipophilicity series and variations in the same.

Supported by NSF (PCM 7620605) and by USPHS (GM 24749)

University of California, Dept. of Physiology, The Center for the Health
Sciences, Los Angeles, Ca. 90024/USA

# New Ion-Selective Membranes

D. AMMANN, D. ERNE, H.-B. JENNY, F. LANTER, W. SIMON

## Introduction

Extensive studies of the molecular parameters of ion-selective antibiotics
(26, 29) and synthetic ion-carriers (1, 23, 25, 26), and model calculations
of the interaction of hard cations with electrically neutral compounds (12,
13) formed the framework for the development of a variety of ion-selective
ligands (11). These have been used successfully as ion-selective components
in liquid membrane electrodes (10, 14). Today, for the potentiometric measurement of the activity of alkali- and alkaline earth metal cations, neutral carrier based electrodes with selectivity for $Li^+$ (4, 28), $Na^+$ (5, 27),
$K^+$ (17, 21), $Ca^{2+}$ (24, 30), and $Ba^{2+}$ (6) are available. These membrane electrodes are now widely used, especially in physiology (7) and medicine (10).
In these areas a $Mg^{2+}$-selective electrode would be of great interest. In the
following, the design features of a $Mg^{2+}$-selective ligand are briefly discussed and a new solvent polymeric membrane electrode with just sufficient
$Mg^{2+}$-selectivity for intracellular studies is presented.

## Design of $Mg^{2+}$-Ionophores

In the search for ligands that in lipophilic membranes behave as ionophores
(ion-carriers) for metal cations, the most important parameters to be considered are: complex stabilities (26), ion exchange kinetics (2), or kinetics of the ion transfer between aqueous and membrane phase (18, 19), and
lipophilicity (20). To obtain high selectivity for a given ion, molecular
parameters such as the number of binding sites and their arrangement (cavity), type of ligand atoms, dipole moments and polarizabilities of the coor-
dinating sites as well as over-all size of the ligand are decisive (13, 23,
26). The choice of these parameters is complicated by the possible formation
of complexes of different stoichiometries and/or by synthetic methods available.

Potential binding sites have to be able to displace water molecules from the
hydration sphere of $Mg^{2+}$. An estimate of the effectiveness of different sites
may be obtained by simple model calculations (13). The calculated (12) energies of interaction in the gas phase of $Na^+$, $K^+$, $Mg^{2+}$, and $Ca^{2+}$ with one
water molecule (Table, column 2) and the corresponding experimental values
(8) for $Na^+$ and $K^+$ (column 3) are in good agreement. In columns 4 to 6 the
increments of the interaction energies are given (relative to column 2)
which result from changes in the dipole moment (column 4), the polarizability (column 5), and the radius of the ligand atom (column 6). The data indicate that an increase in the dipole moment, the polarizability, or a decrease
in the radius of the ligand atom lead to an increased stability of the hypothetical complexes. These effects are especially large for $Mg^{2+}$. Therefore,
polar binding sites (e.g. amide groups (11)), ligand atoms of high polarizability (e.g. N or S (11)), and small ligand atoms generally tend to prefer
$Mg^{2+}$ over $Na^+$, $K^+$, and $Ca^{2+}$.

Table 1. Interaction between binding site models and cations

| ion | interaction energy [kcal/mol] for the complex ion-$H_2O$ | | change in interaction energy [kcal/mol] obtained for | | |
|---|---|---|---|---|---|
| | calculated (12) | experimental (8) | increase in dipole moment by O.5 D | increase in polari- zability by $1\overset{o}{A}^3$ | increase in ligand- atom ra- dius by $0.1\overset{o}{A}$ |
| $Na^+$ | -24.4 | -23.2 | -5.2 | -3.6 | +2.3 |
| $K^+$ | -17.5 | -16.2 | -3.9 | -2.0 | +1.4 |
| $Mg^{2+}$ | -75.4 | | -12.4 | -20.3 | +8.9 |
| $Ca^{2+}$ | -54.1 | | -9.7 | -12.5 | +5.5 |

Based on these assumptions, ligands 1 - 3 were synthesized (Fig. 1). These molecules are expected to form an octahedral coordination sphere in a 1:2- (ligands 1 and 2) or 1:3- (ligand 3) cation:ligand complex which comes close to the optimal coordination geometry for $Mg^{2+}$ (13, 26, 31). Although ligand 1 does form a 1:2-$Mg^{2+}$:ligand complex (15, 22), in membranes it induces a selectivity for $Ca^{2+}$ over $Mg^{2+}$ by a factor of about $10^4$ (Fig. 2). This can be explained by the observed formation of highly lipophilic 1:3-$Ca^{2+}$:ligand complexes (16, 22). Surprisingly, the replacement of the ether oxygen atom

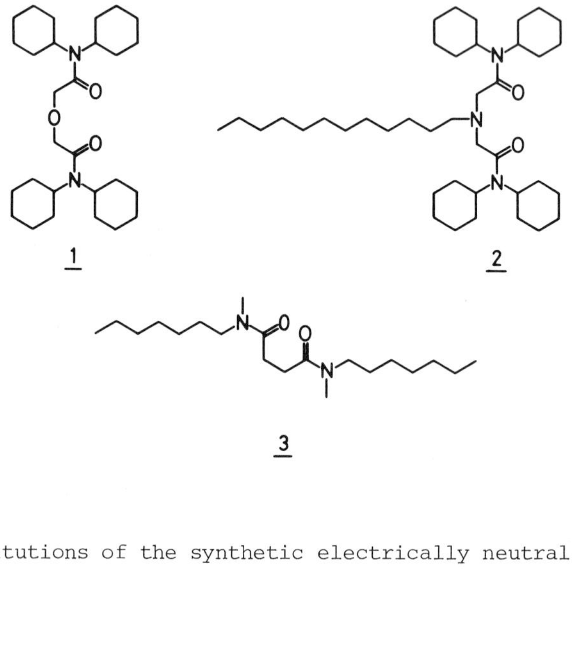

1

2

3

Fig. 1. Constitutions of the synthetic electrically neutral ligands discussed

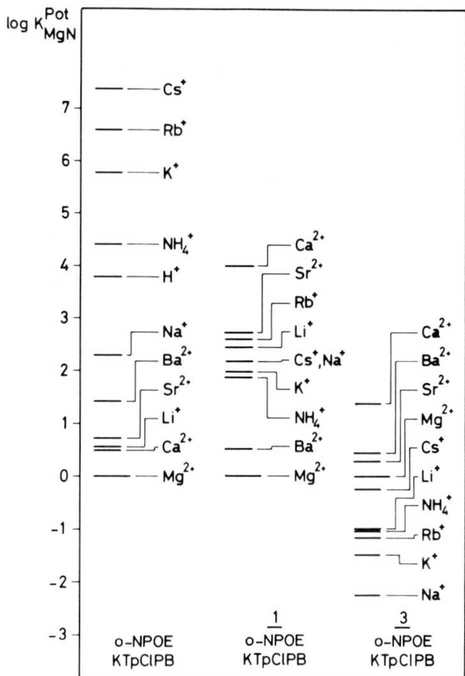

Fig. 2. Selectivity factors, log $K_{MgN}^{Pot}$, of PVC membrane electrodes with o-nitrophenyloctylether (o-NPOE) as membrane solvent. Column 1: ligand free membrane, containing 1wt.-% potassium-tetra(p-chlorophenyl)borate (KTpClPB), 66wt.-% o-NPOE, 33wt.-% PVC. Column 2: 0.2wt.-% ligand 1 (Figure 1), 0.1wt.-% KTpClPB, 69.5wt.-% o-NPOE, 30.2wt.-% PVC. Column 3:1.5wt.-% ligand 3 (Figure 1), 0.8wt.-% KTpClPB, 64.8wt.-% o-NPOE, 32.9wt.-% PVC. The values were obtained by the separate solution method with 0.1M solutions of the chloride salts

(ligand 1) by a tertiary amine nitrogen atom (increased polarizability, ligand 2) does not lead to a preference of $Mg^{2+}$ (3). On account of the high dipole moments of the binding sites, the diamide 3 should be an exemplary $Mg^{2+}$-ionophore (1:3-cation:ligand complex). Indeed in PVC membrane electrodes $Na^+$ and $K^+$ are rejected by a factor of about 100 in respect to $Mg^{2+}$ (Fig. 2). Since $Ca^{2+}$ none the less remains preferred by a factor of about 10, the application of this type of electrode is limited to intracellular $Mg^{2+}$-activity studies.

In Fig. 3 the EMF responses (EMF vs. log $a_{Mg}$) of a cell assembly (3) to pure $MgCl_2$ solutions and to $Mg^{2+}$-solutions containing a representative intracellular electrolyte background are given. In respect of selectivity, it obviously is possible to perform useful $Mg^{2+}$-activity measurements in intracellular environments. The corresponding $Mg^{2+}$-microelectrodes based on ligand 3 have been realized (9).

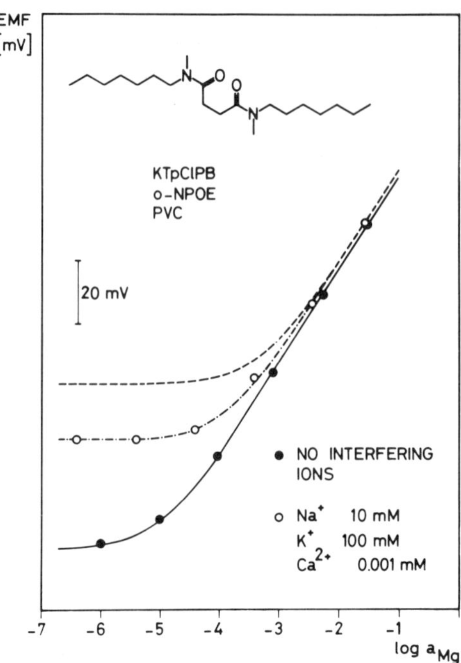



Fig. 3.   EMF response of the PVC membrane electrode based on ligand $\underline{3}$ to different $Mg^{2+}$-activities in the sample solution. Response in pure $MgCl_2$ solutions: lower curve. Response at constant ion background as indicated: middle curve. Response calculated using selectivity data of Fig. 2 and the actual slope of the electrode in solutions with the same ion background: upper curve. Experimental values: dots

## References


1.  Ammann D, Bissig R, Güggi M, Pretsch E, Simon W, Borowitz IJ, Weiss L
    (1975) Preparation of neutral ionophores for alkali and alkaline earth
    metal cations and their application in ion selective membrane elec-
    trodes. Helv Chim Acta 58:1535-1548
2.  Diebler H, Eigen M, Ilgenfritz G, Maass G, Winkler R (1969) Kinetics
    and mechanism of reactions of main group metal ions with biological
    carriers. Pure Appl Chem 20:93-115
3.  Erne D, Stojanac N, Ammann D, Simon W (to be published) Lipophilic
    amides of EDTA, NTA and iminodiacetic acid as ionophores for alkaline
    earth metal cations.
4.  Güggi M, Fiedler U, Pretsch E, Simon W (1975) A lithium ion-selective
    electrode based on a neutral carrier. Anal Lett 8:857-866
5.  Güggi M, Oehme M, Pretsch E, Simon W (1976) Neutraler Ionophor für
    Flüssigmembranelektroden mit hoher Selektivität für Natrium- gegen-
    über Kalium-Ionen. Helv Chim Acta 59:2417-2420

6. Güggi M, Pretsch E, Simon W (1977) A barium ion-selective electrode
   based on the neutral carrier N,N,N',N'-tetraphenyl-3,6,9-trioxaunde-
   cane diamide. Anal Chim Acta 91:107-112
7. Kessler M, Clark LC Jr, Lübbers DW, Silver IA, Simon W (1976) (eds)
   Ion and enzyme electrodes in biology and medicine. Urban & Schwarzenberg, München Berlin Wien
8. Kistenmacher H, Popkie H, Clementi E (1973) Study of the structure of
   molecular complexes. V. Heat of formation for the $Li^+$, $Na^+$, $K^+$, $F^-$,
   and $Cl^-$ ion complexes with a single water molecule. J Chem Phys 59:
   5842-5848
9. Lanter FA, Erne D, Ammann D, Simon W (in preparation) Anal Chem
10. Meier PC, Ammann D, Morf WE, Simon W (1980) Liquid-membrane ion-selec-
    tive electrodes and their biomedical applications. In: Koryta J (ed)
    Medical and biological applications of electrochemical devices.
    John Wiley & Sons, New York Chichester Brisbane Toronto, p 13
11. Morf WE, Ammann D, Bissig R, Pretsch E, Simon W (1979) Cation selectivi-
    ty of neutral macrocyclic and nonmacrocyclic complexing agents in
    membranes. In: Izatt RM, Christensen JJ (eds) Progress in macrocyclic
    chemistry. John Wiley & Sons, New York Chichester Brisbane Toronto,
    p 1
12. Morf WE, Simon W (1971) Berechnung von freien Hydratationsenthalpien
    und Koordinationszahlen für Kationen aus leicht zugänglichen Para-
    metern. Helv Chim Acta 54:794-810
13. Morf WE, Simon W (1971) Abschätzung der Alkali- und Erdalkali-Ionen-
    selektivität von elektrisch neutralen Träger-Antibiotica ("Carrier-
    Antibiotica") und Modellverbindungen. Helv Chim Acta 54:2683-2704
14. Morf WE, Simon W (1978) Ion-selective electrodes based on neutral car-
    riers. In: Freiser H (ed) Ion-selective electrodes in analytical
    chemistry, vol. I. Plenum Press, New York London Washington Boston,
    p 211
15. Neupert-Laves K, Dobler M (to be published)
16. Neupert-Laves K, Dobler M (1978) Crystal structures of metal-ion com-
    plexes with neutral noncyclic ionophores. Abstr. XI. Congr. IUCr,
    p 134
17. Oehme M, Simon W (1976) Microelectrode for potassium ions based on a
    neutral carrier and comparison of its characteristics with a cation
    exchange sensor. Anal Chim Acta 86:21-25
18. Oesch U, Ammann D, Pretsch E, Simon W (1979) Ionophore extrem hoher Lipo-
    philie als selektive Komponenten für Flüssigmembranelektroden. Helv
    Chim Acta 62:2073-2078
19. Oesch U, Simon W (1979) Kinetische Betrachtung der Verteilung von elek-
    trisch neutralen Ionophoren zwischen einer Flüssigmembran und einer
    wässrigen Phase. Helv Chim Acta 62:754-767
2o. Oesch U, Simon W (1980) Life time of neutral carrier based ion-selective
    liquid-membrane electrodes. Anal Chem 52:692-700
21. Osswald HF, Asper R, Dimai W, Simon W (1979) On-line continuous potentio-
    metric measurement of potassium concentration in whole blood during
    open-heart surgery. Clin Chem 25:39-43
22. Pretsch E, Ammann D, Osswald HF, Güggi M, Simon W (1980) Ionophore vom
    Typ der 3-Oxapentandiamide. Helv Chim Acta 63:191-196
23. Pretsch E, Ammann D, Simon W (1974) Design of ion carriers and their
    application in ion selective electrodes. Res Dev 25:20-24
24. Simon W, Ammann D, Oehme M, Morf WE (1978) Calcium-selective electrodes.
    Ann New York Acad Sci 307:52-70
25. Simon W, Morf WE, Ammann D (1977) Calcium ionophores. In: Wasserman RH,
    Corradino RA, Carafoli E, Kretsinger RH, Mac Lennan DH, Siegel FL

    (eds) Calcium-binding proteins and calcium function. North-Holland,
New York Amsterdam Oxford, p 50
26. Simon W, Morf WE, Meier PC (1973) Specifity for alkali and alkaline earth
cations of synthetic and natural organic complexing agents in membranes. Structure and Bonding 16:113-160
27. Steiner RA, Oehme M, Ammann D, Simon W (1979) Neutral carrier sodium
ion-selective microelectrode for intracellular studies. Anal Chem 51:
351-353
28. Thomas RC, Simon W, Oehme M (1975) Lithium accumulation by snail neurones measured by a new $Li^+$-sensitive microelectrode. Nature 258:
754-756
29. Truter MR (1973) Structures of organic complexes with alkali metal ions.
Structure and Bonding 16:71-111
30. Tsien RY, Rink TJ (1980) Calcium-selective microelectrodes are much improved by a new poly(vinylchloride)-gelled sensor. J Neurosci Meth
(in preparation)
31. Williams RJP (1976) Calcium chemistry and its relation to biological
function. In: Duncan CJ (ed) The regulation of intracellular calcium.
Cambridge University Press, England, p 1


## Acknowledgements

This work was partly supported by the Swiss National Science Foundation.


Department of Organic Chemistry, Swiss Federal Institute of Technology,
CH-8092 Zurich, Switzerland


## Discussion

Janata:  My remark relates both to Dr. Eisenman's and Dr. Ammann's lectures.
One implication of Dr. Eisenman's contribution seems to be that there is a
strong effect of coordinated water on the selectivity of the neutral carrier
membranes, and particularly on the anion interference observed with bi-layer
cation permeable membranes. This observation seems to correlate well with
the Hofmeister series for anions. In my opinion, the effect of coordinated
water (as opposed to free water) in the bulk of the membrane (i.e. multi-
component system) has been largely ignored until now. In the light of Dr.
Eisenman's observations it may be of a comparable importance to the primary
ion/ligand interaction.

# Microelectrodes for Novel Anions and their Application to Some Neurophysiological Problems

J.M. PHILLIPS, C. NICHOLSON

The dynamic behavior of the extracellular ions of the brain has been revealed with a new clarity through the use of ion-selective micropipettes (ISMs). Numerous studies have revealed changes in $[K^+]_0$, $[Ca^{2+}]_0$, $[Na^+]_0$ and $[Cl^-]_0$ (see ref. 10 for review). One can go beyond the study of these endogenous brain ions, however, by the introduction of exogenous ionic "probes", and the subsequent monitoring of their behavior with appropriately chosen ion exchangers in ISMs. In this paper we describe the use of anions, novel to the brain, as probes for the diffusion properties of the extracellular space, and as indicators of extracellular volume changes and changes in membrane permeability during spreading depression.

The use of quaternary ammonium cations as probes for diffusion (10, 13, 14) and volume changes (3, 5, 15) has been demonstrated recently because the extracellular microenvironment is a region lined with fixed negative charges, however, (10, 16), it is possible that the behavior of anions in this region differs from that of cations. Hence we have now developed a range of anion probes, together with appropriate ISMs, to enable us to explore the dynamics of anions in the brain cell microenvironment.

The anions that we have used are primarily thiocyanate (SCN), the complex fluorides, hexafluorophosphate ($PF_6$) and hexafluoroarsenate ($AsF_6$) and $\alpha$-naphthalene sulfonate ($\alpha$-NS), ranging in ionic size from 0.42 - 1.12 nm (15). In a few experiments, salicylate (SAL) and hexafluorantimonate ($SbF_6$) anions were also used.

For potentiometric detection of the anion probes, we have employed two different liquid membrane exchangers. The first consists of tricaprylylmethyl ammonium chloride (Aliquat 336), dissolved in 3-nitro-o-xylene at 0.1% v/v. The exchanger is similar to that developed for macroelectrodes by Coetzee and Freiser (2), but with improved selectivity over chloride, the major endogenous anion interferent, by substitution of 3-nitro-o-xylene for 1-decanol as solvent. The 3-nitro-o-xylene probably enhances the electrode selectivity over chloride because the larger and less hydrated anion probes should have a greater partition into it than into 1-decanol, a hydroxilic solvent expected to solvate anions to a degree more nearly equal to that of water. The higher dielectric constant of 3-nitro-o-xylene (about 3.5 times that of 1-decanol) would ensure that relative selectivity would be a strong function of partition coefficient (4).

The second anion exchanger utilized the chloride salt of the dye Crystal Violet as the ionogen, in 3-nitro-o-xylene at 0.02 M. This is similar to that described for macroelectrodes by Ishibashi, Kohara and Horimouchi (6), except their exchanger employed other solvents and a different Crystal Violet concentration.

## ISM Fabrication and Performance

Electrodes were fabricated with double-barrel theta glass tubing. The ion sensitive barrel was silanized at the tip (9) and backfilled with either a 150 mM solution of the sodium salt of the anion to be measured, or a 150 mM probe salt solution diluted by 1/2 with 150 mM NaCl. Although exchangers were previously prepared by repeatedly equilibrating the exchanger with high concentrations of aqueous solutions of the probe (15), it has been found that this step is not necessary for good functioning of the electrode; apparently the exchange of the chloride from either Aliquat or Crystal Violet for the probe anion can take place rapidly when the exchanger is in the electrode exposed to the backfilling solution. After about 30 - 60 minutes of drifting, the potential of such freshly made electrodes is stabilized.

Selectivities of the two exchangers for individual probes were obtained by measuring the electrode potentials in solutions with varying concentrations of probe anions and a fixed background of 150 mM of the interfering ion, then fitting the results to the Nicolsky equation, as previously described (11).

Table 1. Characteristics of anion probe electrodes. Selectivity $K_{ij}$ was determined by a fixed interference method, and fitting the results to the Nicolsky equation, $E = (m)\log([A_i]_o + K_{ij}[I_j]_o)$, where $[A_i]_o$ is the probe concentration, and $[I_j]_o$ is the concentration of the interfering species. The electrode slope, $m$, is the electrode response in millivolts to an activity change of the probe by a factor of 10. Exchangers based on either Aliquat 336 or Crystal Violet displayed similar selectivity characteristics, and typical values for either are presented in the table. The range represents the concentrations of calibrating solutions over which the slope was constant. Because of the very high degree of selectivity for these anions, in the brain one would expect virtually no significant interference from the endogenous anions chloride, bicarbonate, or phosphate

| ANION (i) | INTERFERING ION (j) | $K_{ij}$ X $10^3$ | SLOPE (mV) | RANGE (M) X $10^{-3}$ |
|---|---|---|---|---|
| SCN | Chloride | 1.10 | 54.8 | 0.1 - 10 |
| $PF_6$ | Chloride | 0.35 | 60.6 | 0.1 - 4 |
| $AsF_6$ | Chloride | 0.26 | 60.5 | 0.1 - 5 |
| $\alpha$-NS | Chloride | 0.66 | 60.1 | 0.01 - 5 |
| $AsF_6$ | $HCO_3$ | 0.22 | 60.4 | 0.1 - 4 |
| $AsF_6$ | $PO_4$ | 0.05 | 58.0 | 0.1 - 4 |
| $\alpha$-NS | $HCO_3$ | 0.54 | 61.0 | 0.1 - 4 |

The selectivity coefficients of both exchangers used were similar, and typical values, that applied to either, for the ions SCN, $PF_6$, $AsF_6$ and $\alpha$-NS are shown in Table 1. Not included in the table are the selectivity for SAL (800:1 over chloride) and $SbF_6$ (about 600:1) over chloride). These two probes

were not used as routinely as the other anions, because of the known physiologic effects of SAL, and because of occasional drift and a super-Nerstian
response of the $SbF_6$ electrode. On average, the Crystal Violet based exchanger produced electrodes with less noise and drift, and were thus preferenti-
ally used.


## Diffusion Studies

To investigate anion diffusion either $AsF_6$ or $\alpha$-NS was iontophoresed at some
fixed distance away from an ISM in both 0.3% agar or the cerebellar cortex
of urethane anesthetized rats, as previously described for other probes (13).
The transient increase in anion concentration observed when the iontophoresis current was stepped was Fickian, and for both $AsF_6$ and $\alpha$-NS was in accord
with their expected diffusion coefficients in aqueous medium, indicating that
the electrode response was rapid enough to accurately record transient diffusion. When the same probes were iontophoresed in the molecular layer of
the cerebellar cortex, diffusion obeyed Fick's law, but was altered in a way
consistent with the medium having a tortuosity of about 1.65, and an extracellular volume fraction of about 23 - 25 %. These compare favorably with respective values of 1.57 and 24% for the cation probe (13, 14). Thus, within
the sensitivity of this technique, the diffusion of cation and anion probes
is virtually identical and not affected by surface changes on cells.


## Extracellular Volume and Permeability Changes During Spreading Depression

The dynamic potentialities of the brain cell microenvironment are revealed
during the phenomenon of spreading depression (SD) (1, 8). During this pe-
culiar phenomenon, large amounts of NaCl leaves the extracellular space and
apparently enters the intracellular compartment (7, 12). This should lead to
water movement, cellular swelling, and consequent diminishing of the extracellular volume. Such volume changes and the underlying cellular permeability variations can be monitored directly with ionic probes.

When the anion probe $\alpha$-NS is superfused in a chloride reduced ringer to condition for SD (12) and SD elicited by repetitive stimulation, the $\alpha$-NS
is observed to increase by a factor of 1.5 - 2.0, in a manner similar to
tetraalkylammonium cations (15), suggesting that the $\alpha$-NS remains predominantly extracellular during SD and hence its behavior is reflecting the shrinkage of extracellular volume. When a range of anion probes smaller than $\alpha$-NS
are superfused (or iontophoresed, in the case of SAL), it is seen that some
do not increase in concentration, but actually fall, indicating their entry
into cells (Fig. 1A). By arranging these probes in a sequence corresponding
to their ionic size, an estimate of the size of the apparent channel that
permits movement of anions from the extracellular to the intracellular compartment is revealed (Fig. 1B).


## Conclusion

These studies reveal that the use of probe ions can provide a direct approach
toward the study of certain biophysical phenomena of the extracellular microenvironment. An important finding is the general agreement in the behavior
of cation and anion probes, especially in predicting such physical parameters

as extracellular volume fraction and tortuosity. This tends to support not only the validity of the technique of measuring probe ions, but also the use of extracellular ISMs in general.

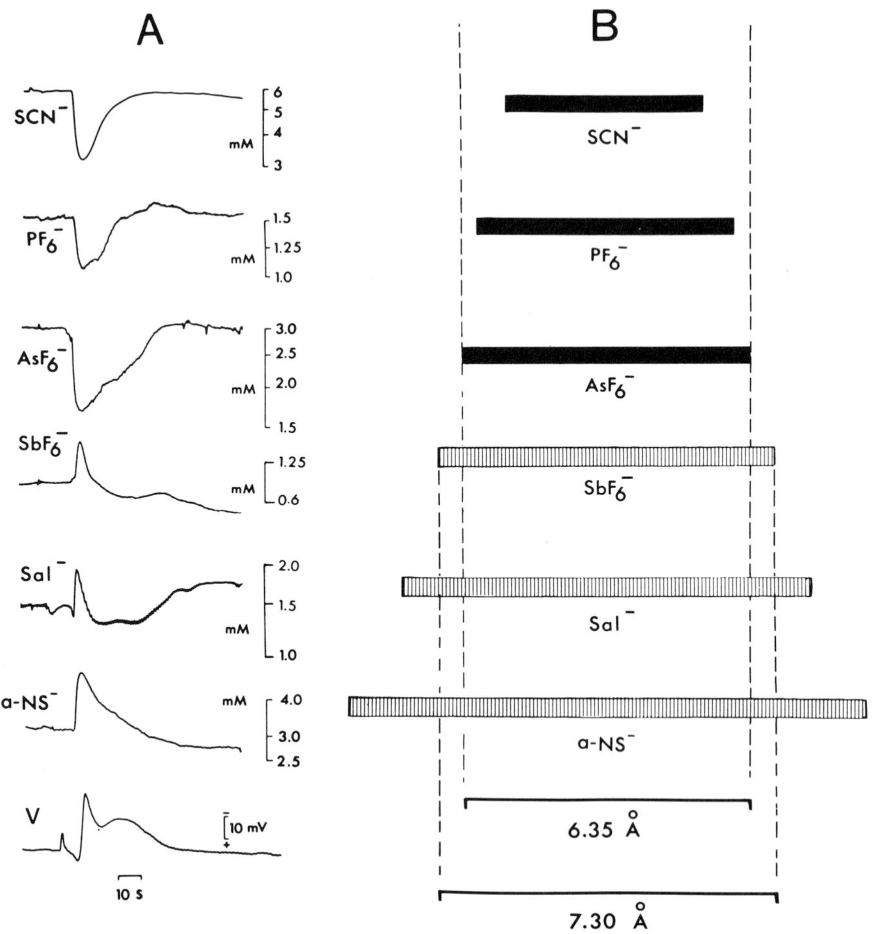



Fig. 1. Behavior of anion probes during SD. A. Recordings with ISMs in the molecular layer of the rat cerebellar cortex. Anion probes have been introduced by superfusion (except for SAL) in a low chloride, SD-conditioning ringer. Salicylate was introduced by iontophoresis from a nearby electrode because it did not easily enter the extracellular space by superfusion across the pial barrier. Each SD was recorded in a separate animal. Anions the size of $AsF_6$ and smaller penetrate cells during SD and decrease in concentration. Anions larger than $AsF_6$ are initially concentrated during SD, indicating relative impermeance of the cells to these species, and reduction of extracellular volume. The SCN, Sal, and $AsF_6$ signals were recorded with Aliquat 336 exchanger, and $PF_6$, $SbF_6$, and α-NS were recorded with the Crystal Violet exchanger. The potential shown (V) is from the α-NS reference barrel, but all ion signals have had their respective DC reference potentials subtracted, and thus represent ion activity only. B. Illustration of the dependence of anion probe behavior during SD on ionic size. The length of each bar is proportio-

nal to ion diameter, SCN being 0.42 nm, $PF_6$ 0.56 nm, $AsF_6$ 0.635 nm, $SbF_6$ 0.73 nm, Sal 0.89 nm, and $\alpha$-NS 1.12 nm. Ions represented by solid bars decrease in extracellular concentration during SD, and those represented by shaded bars initially are concentrated during SD. Thus, an estimate of anion channel diameter is between 0.635 and 0.730 nm.

## References


1.  Bures J, Buresova O, Krivanek J (1974) The mechanism and applications of Leao's spreading depression of electroencephalographic activity. Academic Press, New York
2.  Coetzee CJ, Freiser H (1969) Liquid-liquid membrane electrodes based on ion association extraction systems. Anal Chem 41:1128-1130
3.  Dietzel I, Heinemann U, Hofmeier G, Lux HD (1980) Transient changes in the size of the extracellular space in the sensorimotor cortex of cats in relation to stimulus induced changes in potassium concentrations. Exp Brain Res (in press)
4.  Eisenman G (1968) Similarities and differences between liquid and solid ion exchangers and their usefulness as ion specific electrodes. Anal Chem 40:310-320
5.  Hansen AJ, Olsen CE (1980) Brain extracellular space during spreading depression and ischemia. Acta Physiol Scand 108:355-365
6.  Ishibashi N, Kohara H, Horinouchi K (1973) Aromatic sulfonate ion selective electrode membrane with Crystal Violet as ion exchange site. Talanta 20:867
7.  Kraig RP, Nicholson C (1978) Extracellular ionic variations during spreading depression. Neurosci 3:1045-1059
8.  Leao AAP (1944) Spreading depression of activity in the cerebral cortex. J Neurophysiol 7:359-390
9.  Lux HD (1974) Fast recording ion specific microelectrodes: Their use in pharmacological studies in the CNS. Neuropharmacol 13:509-517
10. Nicholson C (1980) Dynamics of the brain cell microenvironment. Neurosci Res Prog Bull 18:177-322
11. Nicholson C, ten Bruggencate G, Stöckle H, Steinberg R (1978) Calcium and potassium changes in extracellular microenvironment of cat cerebellar cortex. J Neurophysiol 41:1026-1039
12. Nicholson C, Kraig RP (1980) The behavior of extracellular ions during spreading depression. In: Zeuthen T (ed) The application of ion-selective electrodes. Elsevier/North-Holland, Amsterdam
13. Nicholson C, Phillips JM, Gardner-Medwin AR (1979) Diffusion from an iontophoretic point source in the brain: Role of tortuosity and volume fraction. Brain Res 169:580-584
14. Nicholson C, Phillips JM (1979) Diffusion of anions and cations in the extracellular micro-environment of the brain. J Physiol 296:66P
15. Phillips JM, Nicholson C (1979) Anion permeability in spreading depression investigated with ion sensitive microelectrodes. Brain Res 173: 567-571
16. Schmitt FO, Samson FE Jr (1969) Brain cell microenvironment. Neurosci Res Prog Bull 7:277-417


## Acknowledgements

Supported by USPHS Grants NS-13742 and GM-07308 (J.M.P.)

Department of Physiology and Biophysics, New York University Medical Center
550 First Avenue, New York, NY 10016

## Discussion

Kessler:  Did you define or investigate the physiological state of your brain preparation before and during superperfusion, e.g. in terms of micro-circulation, capillary $O_2$ saturation of hemoglobin, local oxygen supply, redox state of respiratory enzymes and activities of physiological ions ($K^+$, $Na^+$, $Ca^{2+}$, $H^+$)?

Nicholson:  We did not monitor these parameters but in all cases the locally evoked cerebellar responses were recorded throughout the experiment and no significant changes were seen in the typical pre- and postsynaptic components of the field potentials. In the superfusion experiments, the probe concentration in the perfusate never exceeded 10 mM and was usually 5 mM. In the iontophoretic experiments, the total amount of substance introduced was extremely small and localized.

Kessler:  The way you induce ion diffusion in your brain preparation by superfusion must cause heterogeneities of many parameters, e.g. ion activities, microflow pattern etc. Did you measure it and take it into consideration when you interpreted the results of your investigations?

Nicholson:  This could be true but experience has generally shown that synaptic transmission in the cerebellum is very sensitive to insult and this continued to be maintained in our experiments.

de Hemptinne:  When I use salicylate as an anion substitute in my experiments I discover that it rapidly enters cells. Why do you therefore see a concentration during SD?

Nicholson:  Of course a weak acid such as salicylate will enter cells in the undissociated form and may lead to measureable pH changes. But the actual quantities necessary to bring this about are quite small and the bulk of the dissociated, ionized anion remains extracellular and is concentrated during SD.

Amman:  Could your method be used with a $Ba^{2+}$-sensitive ISM?

Nicholson:  In principle it could but two factors might make it difficult. Firstly, $Ba^{2+}$ has pronounced neurophysiological effects when applied extracellularly and secondly, it is difficult to consistently iontophorese $Ba^{2+}$ as indeed any divalent ion, from a micropipette - this makes diffusion studies impracticable.

Eisenman:  Do you find it surprising that cations and anions behave so similarly in the extracellular space?

Nicholson:  Yes, we did find it surprising, but this may indicate that the density of fixed negative extracellular charges, per unit volume of extracellular space, is fairly low in the brain.

Acker:  How sensitive is your diffusion method to changes in the distance between the iontophoretic electrode and the measuring electrode?

Nicholson:  The method is rather sensitive to this parameter. Although we take considerable precautions to reduce error, by using parallel electrodes and continuous observation with measuring microscopes, errors of a few microns are unavoidable. Usually we use a spacing of about 100 µm and here an error of 5 - 10 µm would affect the results somewhat. A method of measuring the spacing between electrode tips in the tissue to resolution of a few microns would be most useful.

# Carrier Based Ion-Selective Liquid Membrane Electrodes and their Medical Applications

D. AMMANN, H.-B. JENNY, P. ANKER, U. OESCH, W. SIMON

## Introduction

In clinical diagnosis and therapy the $Na^+$- and $K^+$-concentrations in blood
and urine are important parameters and are generally determined by flame
photometry. During the past few years, however, in clinical laboratories
potentiometric $Na^+$- and $K^+$-measurements with ion-selective elctrodes have
become of increasing importance (4). The use of such membrane electrodes
would be especially attractive alternatives to flame photometry if direct
measurements in whole blood and undiluted urine could be performed. In this
case the potentiometric method would offer advantages such as continuous
monitoring (1, 6, 8) and routine measurements of the ion-activities in
undiluted samples (3, 7). Here we report on the use of such liquid membranes for $Na^+$- and $K^+$-measurements in  undiluted urine samples.

## Potentiometric $Na^+$- and $K^+$-Determinations in Undiluted Urine

It is now recognized that $K^+$-selective antibiotic valinomycin and a $Na^+$-
selective synthetic ligand (see Figs 2 and 3) liquid membrane electrodes,
in respect to selectivity, allow reliable $K^+$- and $Na^+$-concentration studies in blood as well as in urine (4). Membranes of this type are used in a
series of commercial analysers (see Table). From the table it is obvious
that the $K^+$-selective liquid membrane has found wide application for measurements in blood and diluted urine, whereas the $Na^+$-selective liquid membranes seem only rarely to have been used. The fact that neither membrane has
so far been used for analysis in undiluted urine is surprising, however, a
comparison of $K^+$-measurements in undiluted urine samples obtained potentio-
metrically and flame photometrically clearly indicates that the $K^+$-read-
ings for the electrode are too low (Figure 1). A dilution of the urine
samples with O.1 M $MgCl_2$ solutions leads to correct results (Figure 1).
This behaviour cannot be rationalized by assuming the presence of interfering cations. Possible explanations are interferences by lipophilic sample
anions (5), and the presence of complexing agents for potassium in urine.
Even acidification of the urine sample to about pH=2 does not improve the
situation. From this it seems unlikely that the effect is due to complex
formation. That lipophilic anions, here as thiocyanate, interfere is made
evident by Figure 2. Only recently, membranes based on valinomycin showing
drastically reduced anion-interference were characterized (2). Decidedly
improved correlations with flame photometry were obtained (2).

22

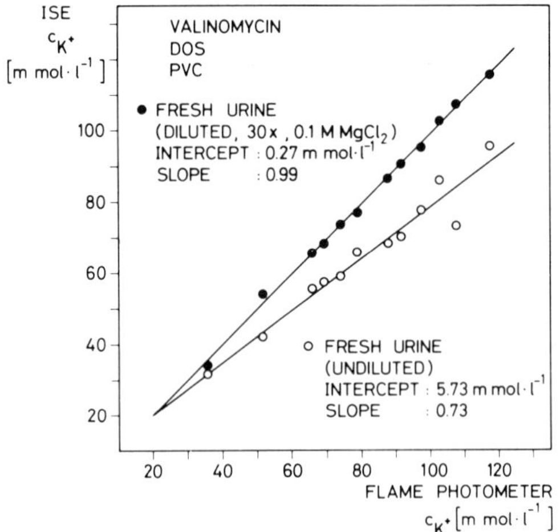

Fig. 1. Correlation of results of K⁺-measurements in diluted (●) as well
as undiluted (o) urine samples obtained by potentiometry (ISE), and by
flame photometry. Membrane composition: 1 wt.-% valinomycin, 66wt.-% bis-
(2-ethyl)-hexyl)sebacate (DOS), 33 wt.-% PVC. Linear regressions give the
solid lines, i.e. $Y = (0.27 \pm 1.25) + (0.99 \pm 0.01) \cdot X$ for diluted samples
and $Y = (5.73 \pm 4.68) + (0.73 \pm 0.06) \cdot X$ for undiluted samples (std. dev.
given). Y represents the potentiometric measurements and X those using
flame photometry. The residual standard deviations of Y and X are $\pm$ 1.16 mM
and $\pm$ 4.35 mM for diluted and undiluted urine samples respectively

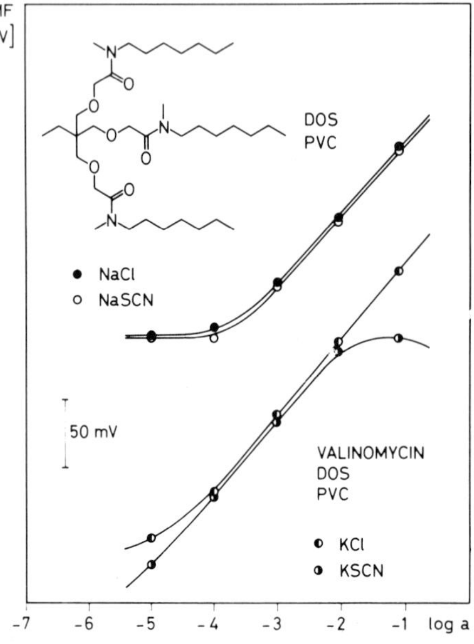

Fig. 2. Response of a Na⁺-selec-
tive (upper) and a K⁺-selective
(lower) PVC membrane electrode to
Na⁺ and K⁺, respectively, in
sample solutions containing anions
of different lipophilicity. Composition of the Na⁺-selective membrane: 5 wt.-% ligand, 63 wt.-%
bis-(2-ethyl-hexyl)sebacate (DOS),
32 wt.-% PVC. For the composition
of the K⁺-selective membrane see
Figure 1

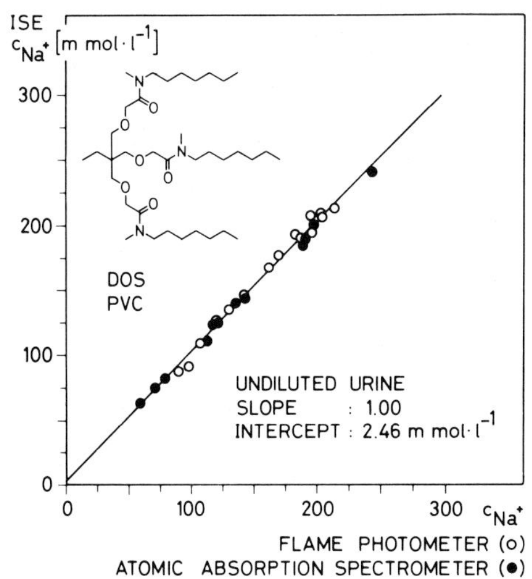


Fig. 3. Correlation of results of Na$^+$-measurements in urine samples obtained by potentiometry (ISE: undiluted samples), and by flame photometry (o) or atomic absorption spectrometry (●). (Membrane composition see Figure 2). A linear regression gives the solid line, i.e. Y = (2.46 $\pm$ 2.76) + (1.00 $\pm$ 0.018) · X (std. dev. given), where Y represents the potentiometric measurements and X those using atomic absorption spectrometry or flame photometry. The residual standard deviation of Y on X is $\pm$ 4.49 mM

Table. Representative examples for the use of Na$^+$- and K$^+$-selective liquid membranes in clinical analyzers (see (4))

| sample | Na$^+$-selective membranes (based on a synthetic ligand) | K$^+$-selective membranes (based on valinomycin) |
|---|---|---|
| blood serum (diluted or udiluted) | HITACHI 702 | ORION SS 30<br>NOVA 1<br>TECHNICON STAT/ION<br>TECHNICON C 800<br>AVL 980<br>HITACHI 702<br>ASTRA 4 / ASTRA 8<br>KODAK |
| urine (diluted) | HITACHI 702 | NOVA 1<br>TECHNICON STAT/ION<br>TECHNICON C 800<br>ASTRA 4 / ASTRA 8 |
| urine (undiluted) | – | – |

For the $Na^+$-selective membrane neither an anion interference (Figure 2)
nor a poor correlation with photometric measurements (Figure 3) on undiluted urine is observed. This supports the notion that the poor correlation
observed for the original $K^+$-selective PVC-membrane as regards measurements in undiluted urine is due to anion interference.

These results indicate that a continuous monitoring of $Na^+$ and $K^+$ in undiluted urine is now within reach.


<u>References</u>


1. Hill JL, Gettes LS, Lynch MR, Hebert NC (1978) Flexible valinomycin
     electrodes for on-line determination of intravascular and myocardial
     $K^+$. Am J Physiol 235: H455
2. Jenny H-B, Riess Ch, Ammann D, Simon W (to be published) Microchim
     Acta
3. Ladenson JH (1977) Direct potentiometric measurement of sodium and
     potassium in whole blood. Clin Chem 23: 1912
4. Meier PC, Ammann D, Morf WE, Simon W (1980) Liquid-membrane ion-selecti-
     ve electrodes and their biomedical applications. In: Koryta J (ed)
     Medical and biological applications of electrochemical devices. John
     Wiley & Sons, p 13
5. Morf WE, Kahr G, Simon W (1974) Reduction of the anion interference in
     neutral carrier liquid-membrane electrodes responsive to cations.
     Anal Letters 7: 9
6. Osswald HF, Asper R, Dimai W, Simon WW (1979) On-line continuous poten-
     tiometric measurement of potassium concentration in whole blood
     during open-heart surgery. Clin Chem 25: 39
7. Preusse CJ, Fuchs C (1979) Klinische Anwendbarkeit ionenselektiver
     Elektroden im Vergleich zur Flammenphotometrie für die $Na^+$- und $K^+$-Be-
     stimmung im Serum. J Clin Chem Clin Biochem 17: 639
8. Treasure T (1978) The application of potassium selective electrodes in
     the intensive care unit. Intens Care Med 4: 83




Swiss Federal Institute of Technology, Dept. of Organic Chemistry, Universitätsstr. 16, CH-8092 Zürich, Switzerland

# Mono-/bivalent Ion Selectivities Obtained by the Nicolsky and the Electrodiffusional Regimes

N. BINDSLEV, A.J. HANSEN

## Introduction

Two methods may be applied for the determination of ion selectivity coefficients. One is the separate solution technique, SST, in which measurements are performed in different solutions containing only one salt. The second is the fixed interference method, FIM, where the solutions contain two salts. The concentration of the interfering ion is kept constant while that of the principal ion varies.

In this communication we wish to issue a warning against an uncritical use of the extended Nicolsky equation in connection with SST. In SST the Nicolsky regime renders two different solutions for the mono-/bivalent selectivity coefficient. Thus the Nicolsky regime does not fulfil the logical requirement of only one value for the selectivity coefficient for monovalent against bivalent ions.

Equations have been derived from the Nernst-Planck relation for electrodiffusion for the mono-/bivalent case. This regime only yields a single solution for the mono-/bivalent selectivity coefficient in SST. We therefore suggest the use of the electrodiffusional regime together with FIM, for the determination of selectivity coefficients.

## Theory

The extended Nicolsky equation is given for two ions, i and j, in equation (1)

$$\Psi = \frac{1}{z_i} \ln \frac{a_i' + K_{ij}(a_j')^{z_i/z_j}}{a_i'' + K_{ij}(a_j'')^{z_i/z_j}} \tag{1}$$

where $\Psi = (V' - V'')F/RT$, $V'$ and $V''$ are the potentials in case ' and ", and $z_i$ and $z_j$ are the valencies of ion i and j, a is the ionic activity and $K_{ij}$ is the selectivity coefficient. It is an impirical relation (1, 2 3).

Eq. (2) gives the total current, $I_{tot}$, in form of the integrated electrodiffusional fluxes for two ions. The constant field assumption has been applied.

$$I_{tot} = -\Psi F \left[ P_i z_i^2 \frac{a_i' - a_i'' \exp(z_i\Psi)}{1 - \exp(z_i\Psi)} + P_j z_j^2 \frac{a_j' - a_j'' \exp(z_j\Psi)}{1 - \exp(z_j\Psi)} \right] \tag{2}$$

$P_i$ and $P_j$ are the permeabilities for ion i and j. The ratio $P_j/P_i$ is equivalent to $K_{ij}$. $I_{tot}$ is zero in potentiometric measurements, i.e.

$$I_{tot} = \sum_{k=1}^{n} z_k F J_k = 0 \tag{3}$$

Here $J_k$ is the flux of ion k. Equations for the mono-/bivalent selectivity coefficients based on either equation (1) or equation (2) and (3) are tabulated in Table 1. In SST the Nicolsky and the electrodiffusional regime give different solutions. Notice that the Nicolsky regime have two different solutions.

**Table 1.**

Separate Solution Technique

| $z_i$ | $z_j$ | Nicolsky $K_{ij} =$ | Electrodiffusion $P_j/P_i =$ |
|---|---|---|---|
| 2 | 1 | $\dfrac{a'_i}{(a''_j)^2\, e^{(2\Delta\Psi)}}$ | $\dfrac{4\, a'_i}{a''_j\, e^{(\Delta\Psi)}\left[e^{(\Delta\Psi)} + 1\right]}$ |
| 1 | 2 | $\dfrac{a'_i\, e^{(-\Delta\Psi)}}{(a''_j)^{1/2}}$ | $\dfrac{a'_i\, e^{(-\Delta\Psi)}\quad e^{(-\Delta\Psi)} + 1}{4\, a''_j}$ |

Fixed Interference Method

| $z_i$ | $z_j$ | Nicolsky $K_{ij} =$ | Electrodiffusion $P_j/P_i =$ |
|---|---|---|---|
| 2 | (1) | $\dfrac{a'_i - a''_i\, e^{(2\Delta\Psi)}}{(a_j)^2\,(e^{(2\Delta\Psi)} - 1)}$ | $\dfrac{a'_i - a''_i\, e^{(2\Delta\Psi)}}{1/4\, a_j\,(e^{(2\Delta\Psi)} - 1)}$ |
| 1 | (2) | $\dfrac{a'_i - a''_i\, e^{(\Delta\Psi)}}{(a_j)^{1/2}\,(e^{(\Delta\Psi)} - 1)}$ | $\dfrac{a'_i - a''_i\, e^{(\Delta\Psi)}}{4\, a_j\,(e^{(\Delta\Psi)} - 1)}$ |

$$\Delta\Psi = (V' - V'')F/RT$$

Derived formulae from the Nicolsky regime (eq. (1)) and the electrodiffusional regime (eqs. (2) and (3)) for both the separate solution and fixed interference methods, SST and FIM. The formulae are only for mono-/bivalent cases. The valencies of the two ions, i and j, are indicated to the left. The ion with a fixed acitivity in FIM is subscripted j.

In SST the Nicolsky regime yields two different solutions, whereas the electrodiffusional regime render two identical solutions, symmetrical. In FIM the exponent is $2\Delta\Psi$ when the activity of the bivalent ion varies and $\Delta\Psi$ with varying activity of the monovalent ion. The two regimes have different solutions in both situations.

V' and V" are the potentials measured with electrodes in case ' and ". The sign convention for $\Delta\Psi$ is shown in Fig. 1.

to the selectivity coefficient whereas the electrodiffusional regime only yields one solution. In FIM the solution for the mono-/bivalent selectivity coefficients depends on which ion is at a fixed activity. In Table I the activity of j is fixed. Also in FIM the two regimes give mutually different solutions. The two regimes have, however, completely identical equations for the mono-/monovalent and the bi-/bivalent cases, not shown.


Internal Correlation. A simple way to check the consistency of the obtained selectivity coefficients is to perform Internal Correlation, in which computed selectivities are compared with actually measured selectivity coefficients, i.e. $K_{ab}$(measured) / $K_{bc}$(measured) $\equiv K_{ac}$(computed) = $K_{ac}$(measured) see Table III.


Material and Methods

Microelectrodes were constructed from rinsed borosilicate glass tubes. The tip-diameter was 1 - 2 $\mu$m. The micropipettes were sialinized for 60 s in dimethylchlorosilane vapor and baked at 100°C for 1 h. The micropipettes were backfilled with 150 mM NaCl. The Na-selective ligand was kindly supplied by Dr. W. Simon (4), ligand No. 3). The ligand was either dissolved in o-Nitrophenyl-ocylether, o-NPOE, 10% (w/w) or in 3-Nitro-O-xylene 10% (w/w). In the latter case the o-NPOE dissolved ligand was rinsed by TCL with a mobile phase of hexane, re-extracted by ether and dryed in a $N_2$ atmosphere. Na-tetraphenylborate was added to both ligand solutions, 0.13% (w/w). A column of 100 - 600 $\mu$m of the ligand solution was sucked into the electrode tip. The only difference in selectivity pattern between the two ligands was a slightly higher selectivity of the 3-Nitro-O-xylene based ligand for Na over Ca. Only results for this type of microelectrode are presented. Potentials were measured on a high-input-impedance potentiometer, $10^{14}$ohm. The reference electrode contained 1M KCl in agar.

Solutions for SST: 150 mM NaCl or KCl and 100 mM $CaCl_2$ or $MgCl_2$ and for FIM: the same as for SST with the addition of one of the other cations as the chloride salt in concentrations 0.1, 1.0 or 10 mM.

Activity coefficients were obtained from Robinson & Stokes (5). In the case of mono-/bivalent ion mixtures the activity coefficients were obtained from Moore & Ross (6), 0.54 $CaCl_2$ and $MgCl_2$ in the concentration range 0.1 - 10 mM with a background of 150 mM naCl or KCl and 0.71 for NaCl and KCl in the concentration range 0.1 - 10 mM with a background of 100 mM $CaCl_2$ or $MgCl_2$.

Fig. 1 shows the difference in sign convention between the Nicolsky, upper panel, and the electrodiffusional regimes, lower panel. The sign convention for the Nicolsky regime is used throughout in Table I.

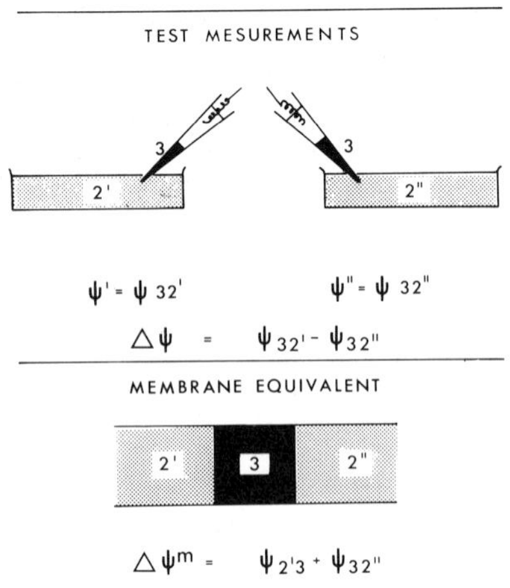



Fig. 1. Sign convention for potential differences in electrode and membrane work. The potential difference measured with electrodes is taken as $\Psi_{32'} - \Psi_{32''}$ between case' and ", upper panel. The same phases as in case' and " in the upper panel is depicted as a membrane equivalent in the lower panel. The potential difference measured across this membrane-equivalent is given by $\Psi_{m'} - \Psi_{m''} = \Delta\Psi^m = \Psi_{2'3} + \Psi_{32''}$. As a result of these conventions the electrode potential difference, $\Delta\Psi$, is equal to the membrane potential difference, $\Delta\Psi^m$, but has the opposite sign, $\Delta\Psi = -\ \Delta\Psi^m$

## Results

Table 2 shows the selectivity coefficients for the Na-selective microelectrode. The obtained selectivities from the two regimes in FIM are fairly consistent. The largest deviation between the two regimes is for Ca/K = 2.3 (97/43). In the

Table 2.

|  | SST | | FIM | |
| --- | --- | --- | --- | --- |
|  | NIC | ELEC. | NIC | ELEC. |
| Na/K | 43 | 43 | 14 | 14 |
| Ca/Mg | 7100 | 7100 | 320 | 320 |
| Ca/Na | 0.3 | 2.9 | 7.0 | 5.5 |
| Ca/K | 1500 | 3300 | 43 | 97 |
| Na/Mg | 7600 | 80 | 52 | 49 |
| K/Mg | 2.1 | 1.0 | 6.3 | 4.9 |

Selectivity coefficients in the separate solution and fixed interference methods, SST and FIM, for a Na-selective microelectrode. There is no general consistency in SST between the Nicolsky (NIC) and the electrodiffusional (ELEC) regimes, whereas these two regimes agree well in FIM. Furthermore there is no general consistency between selectivity coefficients obtained by SST and FIM.

SST, however, there are differences between the selectivity coefficients determined by the Nicolsky and the electrodiffusional regimes of a factor of about 100 (Na/Mg). A comparison of the selectivity coefficients from the SST and the FIM shows that even for the mono-/monovalent and the bi-/bivalent cases (Na/K and Ca/Mg), where the equations in the two regimes are identical, differences exist in the selectivity coefficients measured with the two methods by a factor of up to 22.

Table 3 gives a comparison of measured and calculated selectivity coefficients, Internal Correlation, for both SST and FIM using the Nicolsky and the electrodiffusional regimes. The Internal Correlation is poor in both regimes for SST whereas there is a reasonably good agreement for FIM. The largest deviations are in SST: Na/K (Nicolsky) 4900/43 = 114 and K/Mg (electrodiffusion) 176/1 = 176, and in FIM: Na/K (Nicolsky) 6.2/14 = 0.44 and Ca/Mg (electrodiffsuion) 480/320 = 1.5. The results are for one electrode. Two other electrodes of the same composition gave similar results.

Table 3.

Internal Correlation

| Separate Solution Technique Selectivity Coefficients | | Nicolsky | | Electrodiffusion | |
|---|---|---|---|---|---|
| | | Calc. | Meas. | Calc. | Meas. |
| Na/K | (Ca/K)/(Ca/Na) | 4900 | 43 | 1100 | 43 |
| Ca/Mg | (Ca/K)X(K/Mg) | 3100 | 7100 | 3200 | 7100 |
| Ca/Na | (Ca/mg)/(Na/Mg) | 0.9 | 0.3 | 89 | 2.9 |
| K/Mg | (Na/Mg)/(Na/K) | 176 | 2.1 | 176 | 1.0 |

| Fixed Interference Method Selectivity Coefficients | | Nicolsky | | Electrodiffusion | |
|---|---|---|---|---|---|
| | | Calc. | Meas. | Calc. | Meas. |
| Na/K | (Ca/K)/(Ca/Na) | 6.2 | 14 | 18 | 14 |
| Ca/Mg | (Ca/K)X(K/Mg) | 720 | 320 | 480 | 320 |
| Ca/Na | (Ca/Mg)/(Na/Mg) | 6.1 | 7.0 | 6.5 | 5.5 |
| K/mg | (Na/Mg)/(Na/K) | 3.9 | 6.3 | 3.6 | 4.9 |

Internal Correlation for Nicolsky and electrodiffusional regimes in SST and FIM. Selectivity coefficients for ions a and c were calculated from obtained selectivity coefficients with a third ion, b: $K_{ab}/K_{bc}$   $K_{ac}$ (Calc) and compared with measured values, $K_{ac}$ (Meas).
The Internal Correlation is poor in SST and reasonably good in FIM for both regimes.


## Discussion

The measured electrode potential is the potential created at the interphase between the ion-exchanger and the outer aqueous solution combined with a potential difference between the reference electrode and the electrode tip. The latter, which we call the composite potential, consists of a standard

potential (for the ion-selective electrode), a reference electrode potential and a liquid junction potential. The composite potential may vary with
the ionic strength and the ionic composition of the aqueous solution. In
this study no attempts have been made to differentiate between boundary or
diffusional potentials. On this subject the reader is referred to two excellent papers by Ciani et al.(7) and Morf & Simon.(8).

A mean to avoid variations in the composite potential is to keep the ionic
milieu as constant as possible, e.g. by FIM. The Internal Correlation is
suggested as a simple test of the constancy of this potential, since incon-
sistencies in the Internal Correlation are most likely caused by variation
in the composite potential, see Table III, SST.

There have been much debate about which method to use for a correct determination of the ionic selectivity coefficients for electrodes (9). Use of
the elctrodiffusional regime rather than the Nicolsky is recommended for
the following two reasons. 1) The electrodiffusional regime is based on a
physicochemical mechanism whereas the Nicolsky regime is empirical. 2) the
extended Nicolsky equation yield two solutions for the mono-/bivalent selectivity coefficient in the separate solution technique, SST, where one
solution is the quadrantic form of the other. The logical expectation is
only a single solution for this selectivity coefficient as also yielded by
the electrodiffusional regime.

For biological work an absolute figure for selectivity coefficients is not
mandatory as long as the interference of non-principal ions is neglegible,
but we want to emphasize that a significant interference is only detected
by FIM and not by SST.

In principle it is necessary to have a frame for the comparison of ligands
and in the future to be able to put forward a unifying model for the ionic
discrimination process that also involves the liquid ion exchange electrodes. In conclusion such a model for the ionic discrimination process should
be based on the electrodiffusional description. In the experimental determination of potentials the fixed interference method is recommended as a
mean to keep the composite potential fairly constant. Finally the analysis
of ionic selectivities should involve an Internal Correlation as a simple
test of obtained ionic selectivity coefficients and the reliability of the
data.


<u>References</u>


1. Ciani S, Eisenman G, Szabo G (1969) A theory for the effects of neutral
      carriers such as the macrotetralide actin antibiotics on the electric
      properties of bilayer membranes. J Membr Biol 1: 1
2. Garrels RM, Sato M, Thompson ME, Truesdell AH (1962) Glass electrodes
      sensitive to divalent cations. Science 135: 1045
3. Güggi M, Oehme M, Pretsch E, Simon W (1976) Neutraler Ionophor für
      Flüssigmembranelektroden mit hoher Selektivität für Natrium- gegenü-
      ber Kalium-Ionen. Helv Chim Acta 59: 2417
4. Guilbault GG (1978) Recommendations for publishing manuscripts on
      ion-selective electrodes. IUPAC Inf Bull No 1: 69
5. Lakshminarayanaiah N (1976) Membrane electrodes. Acad Press, London. p
      76 and 124

6. Moore EW, Ross JW (1965) NaCl and $CaCl_2$ activity coefficients in mixed
   aqueous solution. J Appl Physiol 20: 1332
7. Morf WE, Simon W (1978) Ion-selective electrodes based on neutral
   carriers. In: Preiser H (ed) Ion-selective electrodes in analytical
   chemistry, vol 1. Plenum Press, New York-London, Chapter 3
8. Nicolsky BP (1937) The theory of the glass electrode. Zh Fiziol Khim 10:
   495
9. Robinson RA, Stokes RH (1965) Electrolyte solutions. Butterworths,
   London


Department of Medical Physiology, A, The Panum Institute, Copenhagen/Denmark


## Discussion


Buck: The response and selectivity coefficient behavior of $M^+/M^{2+}$ mixtures at a nominally $M^{2+}$ responsive electrode were studied previously by Buck and Sardifer ((1973)J Physiol Chem 77: 2122) and Buck and Stover ((1978) Anal Chem Act 101: 231) Both papers began with formal solutions of the Nernst-Planck equations for internal diffusion potential and the reversible interface of components. The first paper gave approximate numerical solutions, while the second gave digital simulation (exact) solutions for $K_{ij}^{pot}$ vs. $a_i/a_j$. Dependences of $K_{ij}^{pot}$ on $a_i/a_j$ and $u_i/u_j$ (mobility ratios) and on site concentration (membrane loading) were also empirically fit. We showed that no closed-form solutions for response vs. activities are exact and that $K_{ij}(K_{Ca/Na}^{pot})$ is generally activity-dependent. The most-nearly-constant $K_{ij}$ equation has the form:

$$V'' - V' = \frac{RT}{F} \ln \left[ \frac{(a_i')^{1/z_i} + (K_{ij}a_j')^{1/z_j}}{(a_i'')^{1/z_i} + (K_{ij}a_j'')^{1/z_j}} \right]$$

which, for Na/Ca is

$$V'' - V' = \frac{RT}{F} \ln \left[ \frac{(a_{Ca}')^{\frac{1}{2}} + (K_{Ca/Na}\, a_{Na}')}{(a_{Ca}'')^{\frac{1}{2}} + (K_{Ca/Na}\, a_{Na}'')} \right]$$

# Fluorinert Liquids and the Use of Microelectrodes

R. MEYER, W. STOCKEM

Macroplasmodia of the acellular slime mould Physarum polycephalum show oscillating electrical phenomena correlated with periodic protoplasmic shuttle-streaming activity (1). This could be demonstrated by external measurements with microelectrodes performed on specimens which were electrically isolated by the application of fluorinert liquids (FC 43, FC 70 and FC 77 from 3 M-Company, Hilden). FLuorinert liquids are clear, colourless and non-toxic. Light and electron microscopic investigations have shown that the liquids cause no serious structural changes in slime moulds. Single veins of Physarum (see figure) survive in fluorinert liquid for several hours without essential changes in shuttle-streaming and contractile activity.

The physical properties of three fluorinert liquids are listed in the following table:

Table 1.  All values are cited from the producer's publication, except the $O_2$-saturation, which was measured by the authors

| PROPERTY | UNIT | FLUORINERT-LIQUID | | |
| --- | --- | --- | --- | --- |
| | | FC 77 | FC43 | FC 70 |
| NOMINAL  BOILING POINT | °C | 97 | 174 | 215 |
| POUR POINT | °C | -110 | -50 | -25 |
| VAPOR PRESSURE          25 °C | torr | 42 | 1,3 | 0,1 |
| THERMAL CONDUCTIVITY, 25 °C | $\frac{watts}{cm2}$ | 0,055 | 0,058 | 0,061 |
| DENSITY,                25 °C | $\frac{g}{cm3}$ | 1,78 | 1,86 | 1,93 |
| KINEMATIC VISCOSITY,    25 °C | cs | 0,8 | 2,6 | 13,4 |
| VOLUME RESISTIVITY,     25 °C | ohm × cm | $1,9 \times 10^{15}$ | $3,4 \times 10^{15}$ | $2,3 \times 10^{15}$ |
| REFRACTIVE INDEX,       25°C | | 1,28 | 1,291 | 1,303 |
| MOLECULAR WEIGHT | | 415 | 670 | 820 |
| $O_2$-SATURATION, 760 torr atmospheric air 25 °C | $\frac{\mu mol}{ml}$ | 0,261 | 0,268 | 0,282 |

Isolated veins measuring 3 mm in length and 0.5 mm in diameter were mounted in a small chamber filled with FC 43 or FC 70. The measurements were carried out with two glass-microelectrodes inserted into a pair of small water droplets one being placed at either end of the vein. The potential difference between the two droplets was then recorded with conventional amplifier equipment. The procedure of preparing and measuring veins is shown in the following figure:

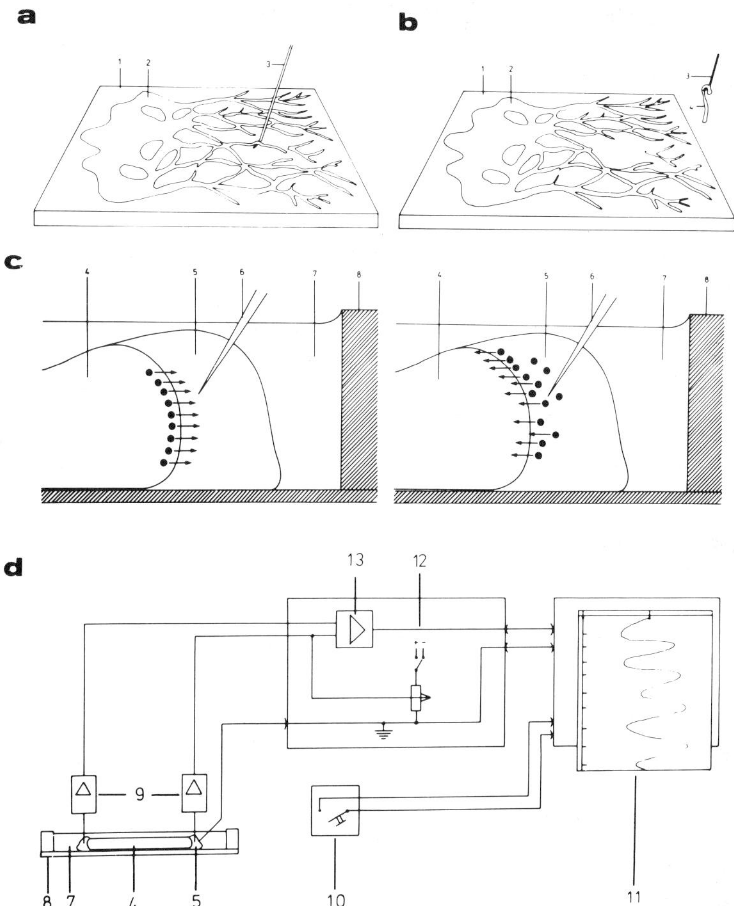

Fig. 1. 1. substrate (agar); 2. plasmodium; 3. needle; 4. separated vein;
5. water droplet; 6. electrode; 7. bathing solution (fluorinert liquid);
8. observation chamber; 9. entrance amplifier; 10. marker to record the
shuttle streaming; 11. pen recorder; 12. compensation device; 13. difference
amplifier.
a) Removal of the vein
b) Air-drying of the vein
c) Hypothetical ion-fluctuations during the measurement
d) Total set up for the experiments
Figure and details from R. Meyer and W. Stockem, 1979


The application of the described method has the adventage over conventional
intracellular measurements with glass-microelectrodes in that the living
organism is prevented from defending itself against the experimental operation by new membrane formation and encapsulation of the tip of the inserted electrode. The results obtained by the fluorinert-liquid-technique could
be confirmed by membrane potential measurements with intracellular glass-
microelectrodes.

## References

1. Meyer R, Stockem W (1979) Studies on microplasmodia of Physarum polyce-
      phalum V: electrical activity of different types of microplasmodia
      and macroplasmodia. Cell Biol Int Rep 3:321-330


Institute for Cytology, University of Bonn, Ulrich-Haberland-Str. 61 a,
5300 Bonn, FRG

# A Double-Channel Ion-Selective Microelectrode with the Possibility of Fluid Ejection for Localization of the Electrode Tip in the Tissue

E. FRÖMTER, M. SIMON[+], B. GEBLER

Microelectrode experiments are often performed as blind punctures without microscopic observation of the impaled tissue. The decision of whether or not a cell was impaled is then based on whether it is possible to record a stable potential difference in the range of the Gaussian distribution of the control values. This approach is of course adequate as long as one is dealing with a homogeneous cell population. However, there are occasions when such a criterion cannot be used, as for example when one tries to puncture subcellular compartments, which may or may not be isopotential with the cell, or when one tries to impale intercellular slits which may be isopotential with the interstitial fluid compartment. In such cases it is imperative to control the impalement by direct microscopic inspection. If the resolving power of the microscope is too low to exactly recognize the electrode tip within the tissue, one can solve this problem by ejecting small amounts of coloured solution from the electrode tip and observing whether or not the colour distributes in the appropriate subcellular or intercellular subcompartment.

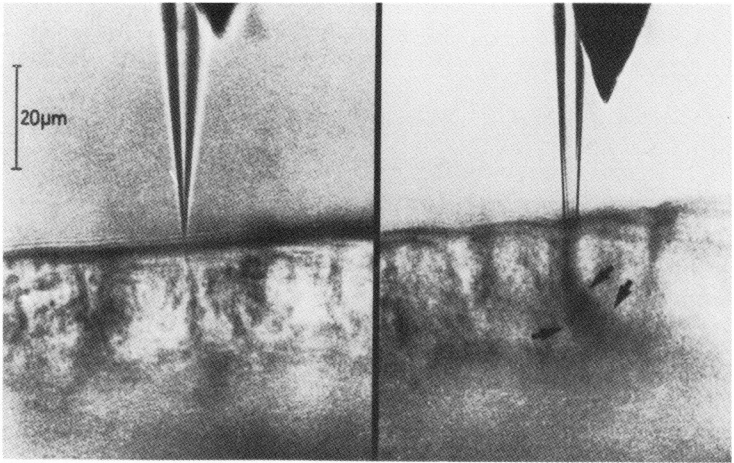

Fig. 1. Microphotograph of a lateral space puncture in Necturus gallbladder epithelium. Left side: electrode tip approaching the mucosal surface of the epithelium near a lateral space. Right side: after successful penetration into the lateral space coloured electrode fluid is ejected from the electrode tip. It flows into the lateral space and outlines its shape (between the arrows). From Curci and Frömter (3)

We have followed this rationale in an investigation (11) of the salt and water flow coupling which is supposed to occur in the lateral intercellular spaces between the epithelial cells of gallbladder epithelium (5, 4). Since this investigation required the determination of the salt concentrations inside the lateral intercellular spaces, we were forced to construct double-channel microelectrodes, one channel of which measured the ion activity (of $Na^+$, $K^+$, or $Cl^-$) and the other channel of which measured the electrical potential and served in addition for localizing the electrode tip by means of fluid ejection. A microphotograph, which demonstrates that fluid ejection can indeed be used to outline the otherwise poorly visible lateral spaces and thus to verify the proper position of the electrode tip during the impalement, is shown in Fig. 1. In the present paper we summarize our experience with the electrode construction.

## I.  Choice of Glass Tubing: Fused Single Glass Barrels versus Theta Glass

Essentially two types of double-channel microelectrodes have been used in classical electrophysiological experiments in the past: 1. microelectrodes pulled from fused single-barrel glass capillary tubing after attaching both tubings and twisting them in the flame by 180 or $360^o$ (7), and 2. microelectrodes pulled from so-called theta glass i.e. capillary tubing into which a flat glass separating wall was inserted to achieve two separate internal channels (2). The first type of double-channel electrodes were also successfully utilized in the past as ion-selective microelectrodes (10) by filling one channel with liquid ion exchanger and the other channel with 3 molar KCl solution (or other appropriate fluids for recording the electrical potential difference). The preparation of double-channel ion-selective microelectrodes from theta glass on the other hand presented so many problems, that thus far, only a few laboratories have succeeded to use such electrodes (6), while most laboratories failed and abandoned this approach.

Table 1.  Comparison of double-channel microelectrodes prepared from fused single capillary tubing and from theta glass

| Electrode cross section: | ∞ | ⊘ |
|---|---|---|
| Construction | easy | more difficult |
| tip size | bigger | finer |
| puncture performance | punctures likely to become leaky | leaky punctures less probable |
| capacitive coupling between channels | small | high |
| electrical connection between channels | less frequent | frequent (depending on technique |

As shown in Table 1, however, double-channel electrodes from theta glass should offer some distinct advantages over electrodes prepared from fused glass capillaries; they should pull to a finer tip size, and their tip should seal better with the cell membrane during the puncture as opposed to the tip of the fused double-barrel electrode type, in which the groove between the barrels, which extends down to the very tip, may be expected to act as a shunt between cytoplasm and interstitial fluid.

One of the major problems with theta glass capillaries has been the channel-to-channel connection. It is often observed that the ion-sensitive liquid flows around the septum and fills the tip of the non-selective channel, or that the liquid retracts inside its channel giving rise to the inflow of reference fluid or bath fluid. Even if these do not occur and both channels appear properly filled under the microscope it is nevertheless often observed that both channels are electrically interconnected. This means both channels respond to changes in external ion concentrations although only one was filled with the ion-sensitive liquid, or channel A responds with a high dc voltage deflection when dc current is passed from channel B into the bath or vice versa.

Table 2.  Possible causes of channel-to-channel connection in double-channel microelectrodes from theta glass

| | | | |
|---|---|---|---|
| 1. | External shunt | | Surface conductance (fluid film) |
| 2. | Internal shunt | a | Incompletely sealed septum |
| | | b | Cracks in septum (after drilling) |
| | | c | Insufficient volume resistivity of the glass material used |
| 3. | Tip connection | a | Septum does not extend to the tip |
| | | b | Protrusion of ion-selective liquid into the reference channel (exchanger capping) |

As shown in Table 2 there are a number of possible causes for such electrical interconnections: 1. external shunts from conductive surface layers on the glass wall; 2. internal shunts from inappropriately sealed or broken separating walls or from insufficient volume resistivity of the glass material and 3. direct connections of both channels at the tip by protrusion of the ion-selective liquid into the opening of the reference channel.

We have tried to overcome those problems by appropriate technical developments. External shunts could be avoided by devising leak-free connections to each glass channel. To exclude internal shunts we have tried to obtain well sealed theta glass. This was a severe problem, since most suppliers sell tubing in which the separating wall is only loosely attached rather than sealed. Suitable glass can be obtained from R & D Optical P.O.B. 198,

Spencerville, Maryland 20868, USA, or from H. Kugelstatter, 12 Römerhofweg, 8046 Garching, Germany. The problem of tip connection has not been resolved completely. However, as a rule tip connection disappears in our electrode after bevelling of the electrode tip.


## Electrode Construction

As shown in Fig. 2 we use theta glass from a Pyrex or Duran 50 brand, with a solid filament attached inside one channel (the prospective ion-selective channel). The glass dimensions are: overall diameter 2 mm, total length: 130 mm, thickness of outer glass wall: 0.25 mm, thickness of separating wall: 0.20 mm, diameter of filament: 0.30 mm. At first a hole is drilled into each channel (see Fig. 2) using a diamond grinding wheel, which is mounted on an electric drill (Minicraft). To assure reproducible positioning of the holes the capillary tubing is mounted in an appropriate holding device during drilling. For drilling the capillary is immersed in water to avoid glass dust. The capillaries are then cleaned in detergent solution (we use any readily available material, usually an anionic detergent) and well rinsed with deionized water. They are then boiled in distilled water for two hours to achieve a reproducible state of hydration. For the sake of convenience

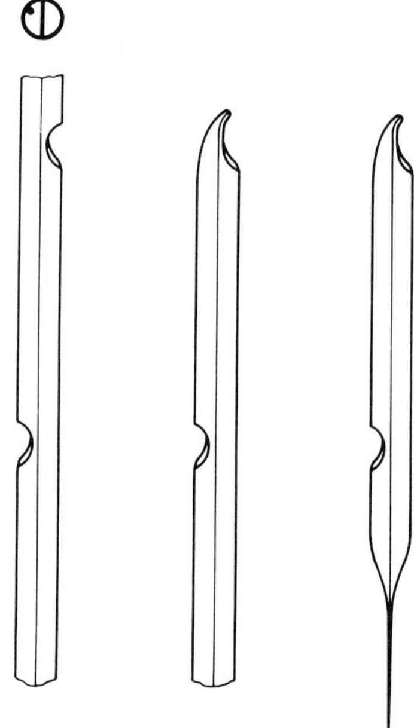

Fig. 2.   Initial steps in the construction of double-channel microelectrodes. Drilling of holes (left), closing the end of the prospective ion-sensitive channel (middle) and pulling the tip (right). We use capillary glass tubing with internal septum (so-called theta glass) which contains a solid filament in the prospective ion-sensitive channel

we use here an apparatus, which allows hot water to circulate both along the
external and internal surface of the glass capillary (Fig. 3). Boiling capillary glass in this apparatus for two or three hours is a standard procedure in our laboratory to reduce the occurrence of tip potentials in normal
KCl filled Ling-Gerard microelectrodes. After boiling, the capillaries are
dried in an oven for 15 min at 200°C. This procedure of hydrating and drying

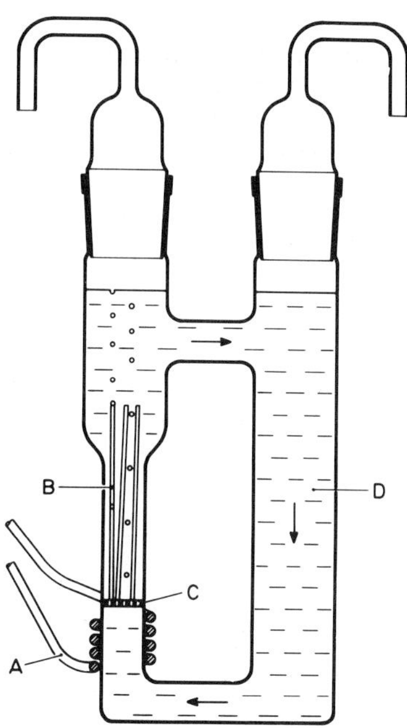

Fig. 3.  Apparatus for hydration of the glass capillary wall. A: heating
cable (Electrothermal HC 102, London, England), B: glass capillaries, C:
porous sintered glass disk supporting the capillaries and allowing the fluid
to circulate as indicated by arrows, D: filling solution is pure distilled
water. The apparatus allows water to continuously circulate along the outer
and inner capillary wall. Besides for the present purpose, this apparatus is
routinely used in our laboratory for the preparation of normal Ling-Gerard
microelectrodes. Our experience is that boiling capillaries for two hours in
this apparatus, drying them briefly for 15 min, pulling and filling them immediately with methanol and later 3 molar KCl solution, prevents the occurrence of tip potentials in normal high resistance (fine-tip) Ling-Gerard
microelectrodes

helps to assure a reproducible state of hydration of the glass surface before silanization. After drying, the upper end of the prospective ion-selective channel is sealed by pulling off the end in a flame (Fig. 2, middle).
Then the tip is pulled in a vertical puller (Narishige) to an overall length
of 17 to 22 mm with a corresponding tip size of approximately 0.3 μm or less.
After pulling, the bottom end of the electrodes is closed temporarily by cap-
ping it with shrink tubing. The electrodes are then tightly mounted (with
the help of plasticine) tip-up in a plexiglass holder and exposed for 60 s
to dimethyl-dichloro-silane vapour in a glass jar (Fig. 4). The vapour enters
the prospective ion-selective channel across the side-hole and silanizes the

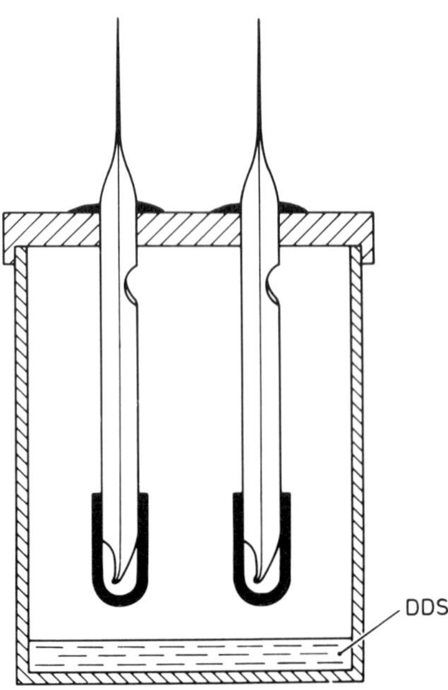


Fig. 4. Technique of vapour silanization. DDS: dimethyl-dichloro-silane. For explanation see text

inside of this channel. Exposure time to the silane vapour is critical. It depends on the length and on the diameter of the channel. To avoid or reduce contamination of the reference channel by silane vapour which might possibly emanate from the electrode tip, a draft of fresh air is blown against the tips during the silanization process. The dimethyl-dichloro-silane liquid must be fresh. If it produces a smoke when exposed to the open air, it can no longer be used. After silanization, the caps are removed and the electrodes are baked for 60 min at $130^{\circ}$ C in an oven. After baking, the ion-selective liquid is filled into the silanized channel from the side-hole via a small catheter. The liquid flows into the tip along the filament and fills it completely overnight. The next morning coloured Ringers solution (1% Kitonblue in frog Ringers) or any other fluid suitable for potential recording is filled into the non-silanized channel from the end via a Hamilton syringe and an appropriate reference fluid is injected into the ion-selective barrel to establish contact with the ion-selective liquid (Fig. 5). The electrical connection to the selective channel is established via an Ag/AgCl wire which is cemented into the side-hole together with a polyethylene tubing for pressure application. Contact with the non-selective channel is achieved by inserting the blunt end of the electrode into a suitable holder. We use a holder similar to that described by Grundfest et al. (8) which allows potential monitoring and simultaneous pressure application; however, instead of a silver wire it contains a calomel half

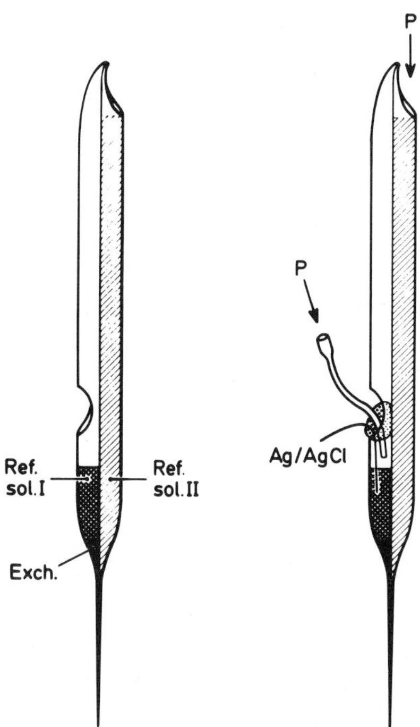



Fig. 5. Final steps in the construction of double-channel ion-sensitive microelectrodes. Exch.: ion-sensitive liquid. Reference solution I establishes the contact between the ion-sensitive liquid and the Ag/AgCl wire electrode. Reference solution II fills the non-selective channel and serves to record the electrical potential difference. By mounting the electrode firmly into an appropriate holder, pressure (P) can be applied to the non-selective channel to eject coloured reference solution and via a small catheter cemented into the side-hole, pressure can be also applied to the selective channel. The latter was sometimes necessary to prevent backflow of the ion-selective liquid which happened to occur occasionally during pressure application to the non-selective channel

cell, and instead of oil it is filled with electrolyte solution. To allow visible ejection of fluid the electrode tip is bevelled on a rotating grinding machine similar to that described by Brown and Flaming (1). Bevelling is performed under visual control to detect the appearance of coloured fluid and under continuous monitoring of the resistance of the non-selective channel. Fluid ejection is probed with a pressure of 5 to 10 atmospheres. Usually sufficient fluid ejection is obtained when the electrode resistance falls to approximately 80 M$\Omega$. The tip diameter is then approximately 0.7 to 1.2 $\mu$m. To avoid dirt in the filling solution which interferes with proper filling and blocks the tips during ejection, we perform most of the work in a dust-free cabinet. The individual steps of the electrode construction are summarized in short form in Table 3.

42

Table 3.  Short summary of preparation technique

 1.  Grind holes in theta tubing

 2.  Clean tubing with detergent, rinse with distilled water

 3.  Boil 2 hrs in distilled water

 4.  Dry 15 min at $200^{\circ}$ C

 5.  Close upper end with flame

 6.  Pull tip

 7.  Cap bottom end and silanize selective channel in DDS* vapour for 60 s

 8.  Remove cap, bake for 60 min at $130^{\circ}$ C

 9.  Fill ion-selective fluid into silanized channel from side

10.  Store overnight in air

11.  Fill reference channel from end and store in Ringers solution

12.  Bevel tip

*DDS: dimethyl-dichloro-silane


## Electrode Performance and Yield

Ion-selective electrodes were produced for $Na^+$, $K^+$ and $Cl^-$ by filling the silanized channel with a $Na^+$ selective cocktail based on a neutral $Na^+$-ligand which was developed by Simon and collaborators (11), or with the liquid ion exchangers Corning No. 477317 or Corning No. 477315, respectively. The $Na^+$ selective cocktail was obtained from Science Trading, 66 Glauburgstr., 6000 Frankfurt, Germany.

After filling, all electrodes were inspected under a Zeiss standard microscope, which was equipped with an Epiplan 80/0.95 (D=0) objective giving an excellent view of the tip at 800 times magnification without requiring immersion. If the above protocol had been followed meticulously, inspection of the electrode tips before bevelling virtually never showed any sign of retraction of the ion-selective liquid along the silanized channel, nor any sign of overflow of the ion-selective liquid into the reference channel. After bevelling, however, in some cases retraction of the ion-selective liquid was noticed, which was probably related to insufficient silanization resulting from the use of old silane fluid. Since bevelling was mandatory to obtain visible fluid ejection only a few electrodes were tested before bevelling. In those electrodes it was observed that the reference channel often exhibited a significant potential response to changes in outside ion concentrations. This was particularly true of the $K^+$ electrodes, which could have a slope of as high as 59 mV per decade activity change with a resistance of 300 to 500 M$\Omega$. This ion-sensitivity of the reference channel, however disappeared as a rule after bevelling and approached the values of the free liquid junction potential between reference fluid and calibration fluid. If the slope of the ion-selective channel was not ideal before bevelling, it increased usually towards the ideal slope after bevelling. The yield of good electrodes having slopes of greater than 50 mV/decade, as tested in pure salt solutions, varied between 30% and 50% in the case of $Na^+$ and $K^+$ electrodes but was only approximately half as high in case of $Cl^-$ electrodes.

Besides the resistance of the reference channel also the resistance of the ion-selective channel was measured before and after bevelling. After bevelling, the resistance was: $K^+$ electrodes: 0.8 to 1.8 $\cdot 10^{10}$ $\Omega$, $Na^+$ electrodes: 0.8 to 1.8 $\cdot 10^{11}$ $\Omega$ and $Cl^-$ electrodes: 0.9 to 4.1 $\cdot 10^{11}$ $\Omega$. Interestingly, bevelling reduced the resistance of both channels in all electrodes in approximately the same proportion (by $\sim$ 3.3 fold). This observation suggests that the electrical interconnection between selective and reference channel (ion-sensitive response of the reference channel) which was observed before bevelling cannot result from insufficient volume resistivity of the separating glass wall in the tip, since in that case the $Na^+$ and $Cl^-$ electrodes should be more affected, because their resistance was higher. In fact, however, the ion-sensitivity of the reference channel (before bevelling) was most pronounced in $K^+$ electrodes, the resistance of which was one order of magnitude smaller.


Conclusion

In the present paper we report a technique to construct double-channel ion-sensitive microelectrodes for $Na^+$, $K^+$ and $Cl^-$ from glass capillary tubing with an internal septum (so-called theta glass). The yield of the technique is at present in the order of 30%. We bevel the tips of these electrodes to an outer diameter of 0.7 to 1.2 um in order to be able to eject coloured reference solution from the electrode tip which helps us to localize the tip during the puncture. We have applied these electrodes to determine the ion concentrations in the lateral intercellular spaces between epithelial cells, which are thought to be the main site of ion and water flux coupling in epithelia. By applying less bevel, electrodes with finer tips can be prepared which should be appropriate for measurements of ion concentrations in small fragile cells.


References


1. Brown KT, Flaming DG (1974) Bevelling of fine micropipette electrodes by a rapid precision method. Science 185:693-695
2. Brown KT, Flaming DG (1977) New microelectrode techniques for intracellular work in small cells. Neurosci 2:813-827
3. Curci S, Frömter E (1979) Micropuncture of lateral intercellular spaces of Necturus gallbladder to determine space fluid $K^+$ concentration. Nature 278:355-357
4. Diamond JM (1979) Osmotic water flow in leaky epithelia. J Membr Biol 51:195-216
5. Diamond JM, Bossert WH (1967) Standing-gradient osmotic flow. A mechanism for coupling of water and solute transport in epithelia. J Gen Physiol 50:2061-2083
6. Dufau E, Acker H, Sylvester D (1980) Double-barrel ion-sensitive microelectrodes with extra thin tip diameters for intracellular measurements. Med Prog Technol 7:35-39
7. Frank K, Becker MC (1964) Microelectrodes for recording and stimulation. In: Nastuk WL (ed) Physical techniques in biological research Vol. V, Part A, Academic Press, New York, pp 22-87

8.  Grundfest H, Kao CY, Altamirano M (1954) Bioelectric effects of ions
    microinjected into the giant axon of loligo. J Gen Physiol 38:245-
    282
9.  Güggi M, Oehme M, Pretsch E, Simon W (1976) Neutral ionophor for liquid
    membrane electrodes of high selectivity for sodium over potassium
    ions. Helv Chim Acta 59:2417-2420
10. Khuri RN (1976) Microelectrodes utilizing glass and liquid ion-exchan-
    ger sensors. In: Kessler M, Clark LC Jr, Lübbers DW, Silver IA,
    Simon W (eds) Ion and enzyme electrodes in biology and medicine.
    Urban & Schwarzenberg, München, pp 131-135
11. Simon M, Curci S, Gebler B, Frömter E (1980) Attempts to determine ion
    concentrations in the lateral spaces between the cells of Necturus
    gallbladder epithelium with microelectrodes. In: Ussing HH, Bindslev N,
    Lassen NA, Sten-Knudsen O (eds) Water transport across epithelia
    (A. Benzon Symposium 15) Munksgaard, Copenhagen (in press)

Max-Planck-Institut für Biophysik, Kennedyallee 70, 6000 Frankfurt (Main),
Germany

[+]Department of Pathophysiology, Medical Academy, Poznan, Poland


Discussion

Galey:  Dr. Frömter, I think this is an exciting approach to testing the
standing gradient hypothesis. However, I do have a question. Since the electrode entry into the L.I.S. of the epithelia takes place through the cells
of the epithelial sheet itself, how can you be sure that the activities you
measure are not significantly influenced by leak pathways created by the penetrating electrode?
Frömter:  This is indeed a great problem, and we are not definitely sure,
whether we have completely solved it. We take the following precautions and
checks to prevent or detect leaky impalements:
1. As reported in the paper, we have chosen theta-glass instead of fused
double-barrel glass tubing for our electrode in order to reduce the possibility of leak artifacts.
2. We control whether we obtain acceptable cell potentials upon retraction
of the electrode tip from the lateral space into the cell and this is what
we normally find.
3. We pass constant current pulses across the tissue continuously to record
the resistance between the microelectrode and the luminal bath and only se-
lect measurements in which the resistance pulses remain constant for at
least 1 min, and which fall into the range of control values, which we have
observed when puncturing with finer single-channel electrodes previously.
4. We reject all measurements with visible leak of coloured fluid from the
lateral space into the lumen.
5. In the case of substantial leaks, we should not only have diffusion of
cytoplasmic ions into the space, but also flow of colour into the cells.
This happens unfortunately all too often, but it is readily recognized and
the respective measurements are discarded.

# A Low-Noise, Rapid Microelectrode

A. HAEMMERLI, J. JANATA, H. MACK BROWN

## Introduction

Ion-selective microelectrodes ($\mu$-ISE) using an ion-selective liquid ion-exchanger membrane have very high resistances because of their very small tip diameter (0.5 - 1.0 $\mu$m). The value of the resistance ranges from $10^9 \Omega$ to $10^{11} \Omega$ (2, 8, 10) depending on the tip size and the type of ion exchanger. This imposes the requirement of a very high input impedance and a very small leakage current on the measuring electrometer.

Any shunt capacitance across the $\mu$-ISE such as shielding of the lead will increase the response time of the probe. On the other hand, shielding is necessary in order to reduce the influence of the electromagnetic noise. It is generally most desirable to bring the electrometer as close as possible to the microelectrode. One possibility is to use a preamplifier which can be placed only a few centimeters away from the $\mu$-ISE. Another possibility is a microprobe electrometer in which the preamplifier is actually part of the $\mu$-ISE's holder. A further step is to place the microtip directly on the gate of the amplifier, which is the subject of this communication.

This article describes the construction of a micro potassium ion-selective field effect transistor ($K^+$ $\mu$-ISFET) and the performance characteristics (sensitivity, time response, selectivity, sensitivity to noise) of this device. The time response of the $K^+$ $\mu$-ISFET is compared to the time response of the potassium selective $\mu$-ISE ($K^+$ $\mu$-ISE). Finally a comparison between the $K^+$ $\mu$-ISFET and the $K^+$ $\mu$-ISE is presented in both intracellular and extracellular cases. Construction details have been recently published (3); Fig. 1 shows a diagram of a $K^+$ $\mu$-ISFET described in this paper. Potassium ion microelectrodes were prepared as described previously (9). The theory of operation of the ISFET has been reviewed in (4).

It is important to note that the pH ISFET used in the fabrication of the $K^+$ $\mu$-ISFET can be reused many times. This is done simply by melting the sealing wax and inserting a new microtip.

After some practice the fabrication of the $K^+$ $\mu$-ISFET, (assuming that we start with a pH ISFET and an empty silanized microelectrode) takes from 10 to 15 minutes.

## Results

### Initial Testing of the Device

When used for the first time, the pH ISFET is allowed to hydrate in a 0.01 M KCl solution while it is run in feedback mode (constant drain current $I_D$) for approximately two hours. To close the circuit an Ag/AgCl (saturated KCl) reference electrode is used. When the device is hydrated its drain current $I_D$ is measured as a function of the gate voltage $V_G$. The leakage current through the whole package is also measured to test the integrity of the device. When a $K^+$ u-ISFET is completed, the $V_G$ - $I_D$ characteristic is measured again and should be the same as that of the bare gate device.

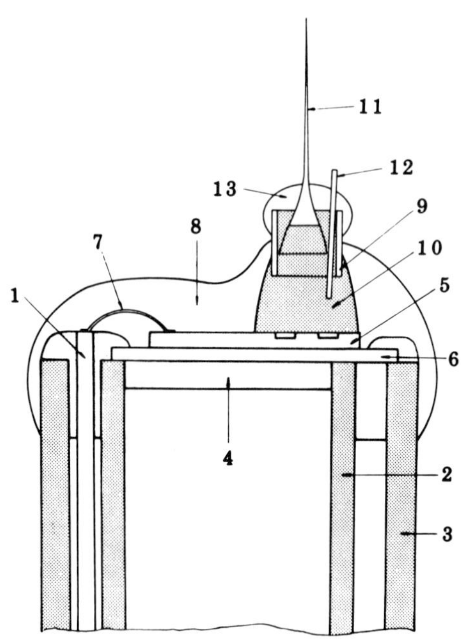

Fig. 1.  Schematic diagram of the $K^+$ u-ISFET. (1) Lacquered copper wires; (2) and (3) glass tubing, 3 mm o.d. and 6 mm o.d., respectively; (4) and (8) epoxy encapsulant; (5) transistor chip; (6) Kovar substrate (7) Al-Si wires; (9) glass tube; (10) buffered polyacrylamide gel, pH 7.0; (11) $K^+$ microelectrode tip; (12) 50 u pressure equalizing capillary; (13) sealing wax

## Measurement of Sensitivity

In order to determine their sensitivities, the $K^+$ u-ISFETs were tested in five solutions containing different concentrations of potassium ($10^{-4}$ M, $10^{-3}$ M, $10^{-2}$ M, 0.1 M and 0.5 M KCl). All solutions had a background of 0.1 M NaCl. The sensitivity (50 $\pm$ 1 mV/log $a_K+$) and selectivity constant (2.5 x $10^{-2}$) were found to be in agreement with values published previously (9, 11).

## Time Response Measurements

The time response of the $K^+$ u-ISFET to a step change in the reference electrode potential was first measured and compared to that of a $K^+$ u-ISE. The comparison was carried out under two different conditions. First the u-ISE was connected to the gate of a MOSFET through an uninsulated copper wire,

then the copper wire was replaced by a shielded cable having a capacitance
C of 145 pF the shield of which was grounded.

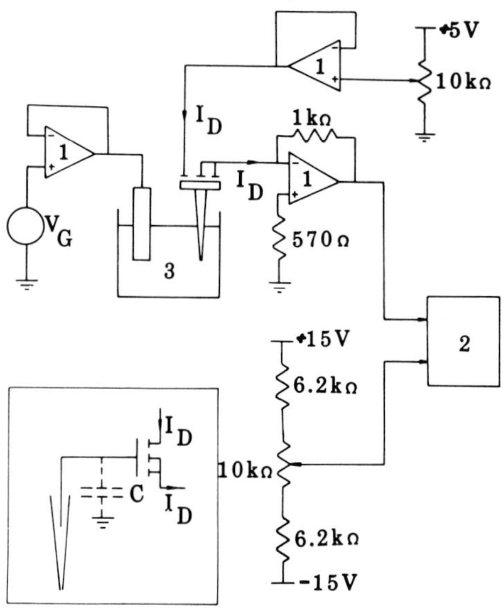


Fig. 2. Arrangement used for time response measurements. (1) UA741 operati-
onal amplifier; (2) storage oscilloscope; (3) O.01 M KCl solution; Inset:
Measurement with /u-ISE

The arrangement used to measure the time response is schematically shown in
Fig. 2. A square wave provided by a function generator (Tektronix model FG
501) is applied at the reference electrode. The signal has a suitable offset
so that there is always a drain current $I_D$ flowing through the ISFET (or the
MOSFET for measurements with $K^+$ /u-ISE). The frequency of the square wave is
set at O.1 Hz. The drain voltage $V_D$ is kept at 2.O V for all measurements.
$I_D$ is fed into a current follower. The output of the current follower and an
adjustable offset voltage are connected to the differential amplifier (Tektronix model 7A22) of a storage oscilloscope (Tektronix model 7623). The accuracy of a single time response measurement varies from $\pm$ 5 ms to $\pm$ 20 ms
depending on the oscilloscope time scale imposed by the value of the time
response.

The time response was measured for both the $K^+$ /u-ISFET (five devices and the
$K^+$ /u-ISE (ten devices); the measurements for the $K^+$ /u-ISE were made in a
Faraday cage to limit noise interference.

The values for the time constant $\tau$ (63% of the full response) for both a step
increase ($\tau_i$) and a step decrease ($\tau_d$) in potential are given in Table 1. It
can be seen that in all three types of measurements $\tau_i$ is shorter than $\tau_d$.
It has been shown earlier for neutral carrier membrane electrodes (8) that
the time constant for a step increase in activity is smaller than for a step
decrease. It is reasonable to expect the same behavior for the electrical

response of the membrane used in our experiment, even though it contains an electrically charged carrier agent. The $K^+$/u-ISFET and the $K^+$/u-ISE have similar time responses when no shielded cable is used. With a shielded cable, the time responses of the $K^+$/u-ISE increased as expected.

Table 1.

| | $\tau_i$ (msec) | $\tau_d$ (msec) |
|---|---|---|
| $K^+$/u-ISFET | 104 $\pm$ 42 | 498 $\pm$ 270 |
| $K^+$/u-ISE (copper wire) | 148 $\pm$ 25 | 485 $\pm$ 94 |
| $K^+$/u-ISE (shielded cable) | 370 $\pm$ 71 | 1660 $\pm$ 515 |

Application

In order to compare the performance of the $K^+$/u-ISFET with that of the $K^+$/u-ISE two experiments were conducted on the large photoreceptors (100 /um) from the barnacle Balanus eburneus (9).

In the experiment with the $K^+$/u-ISE, the membrane potential ($E_M$) measuring electrode (3 M KCl filled microelectrode) and the $K^+$/u-ISE ($E_M + E_K$) were inserted into the cell. A reference electrode (3 M KCl filled broken tip microelectrode) was placed into the extracellular bath. $E_K$ was obtained by feeding the preamplified $E_M$ and ($E_M + E_K$) signals into a differential amplifier.

In the experiment with the $K^+$/u-ISFET, $E_M$ was measured differentially with two potential measuring electrodes, one of them being in the cell and the other one in the extracellular bath. The $K^+$/u-ISFET was introduced into the cell and a reference electrode was used in the extracellular bath. $E_M$ had to be measured differentially because the $K^+$/u-ISFET was operated in feedback mode which implies that the potential applied at the reference electrode was not constant. The cell used with the $K^+$/u-ISFET was different from that used with the $K^+$/u-ISE. The calibration of the devices showed very similar sensitivities both before and after the experiment.

In the first experiment, the cell was bathing in a low $a_K$ extracellular bath. Then the extracellular $a_K$ ($a_K^o$) was raised and the changes in $E_M$ and $E_K$ were monitored. Fig. 3 (intracellular) shows that such an increase in $a_K$ produces a depolarization of the cell ($E_M$ increases). Also, we can see that both the $K^+$/u-ISE and the $K^+$/u-ISFET show an increase in $E_K$ which reflects an increase in intracellular potassium activity ($a_K^i$). The $K^+$/u-ISFET gives a much better quality recording than the $K^+$/u-ISE.

In the second experiment, the $K^+$ selective device was removed from the cell and put in the extracellular fluid. The device was initially in a low $a_K^o$ bath. Then $a_K^o$ was raised in steps and $E_K$ was monitored. Fig. 3 (extracellu-

lar) shows that both devices recorded similar quantitative changes as was
expected since they had similar sensitivities and since the activity changes
were the same. Here again, the difference in quality of the recordings is
striking. This application shows the suitability of the $K^+$ $\mu$-ISFET for intracellular measurements. Quantitative information about barnacle photoreceptors is given in (1).

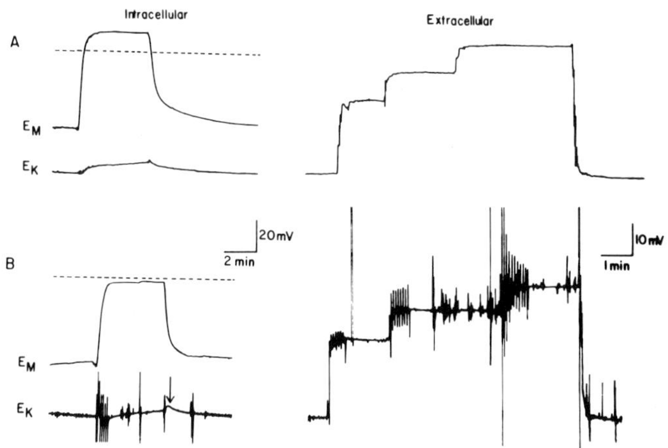


Fig. 3. Physiological application. (A) $\mu$-ISFET recordings. (B) $\mu$-ISE recordings. The zero reference for $E_M$ measurements is indicated by the dashed
line. The arrow indicates an electrical artifact which was due to a slight
imbalance in our differential amplifier

## Discussion

The noise and the speed of a microelectrode are interrelated phenomena. Results in Table 1 show clearly that the response time of the device is increased by the shielding. On the other hand unshielded microelectrodes suf-
fer from the noise interference (Fig. 3).

The equivalent electrical circuit of a liquid ion exchange microelectrode
lead and preamplifier is shown in Fig. 4. It has been shown (10) that the
time constant of a $Si_3N_4$ ISFET is on the order of $10^{-6}$ sec. The value of
$C_{OX}$, $C_D$ therefore does not need to be considered in this case. The depth of
immersion is clearly arbitrary and depends on the particular experiment. The
section of the model between points A and B designated as "LEAD" represents
a simple transmission line from which the voltage $V_{AB}$ (noise) is given (11)
by:

$$V_{AB} = V^+ \exp\left[-(j\omega RC)^{1/2}Z\right] + V^- \exp\left[(j\omega RC)^{1/2}Z\right]$$

where $V^+$ and $V^-$ are complex amplitudes of the traveling wave, $\omega$ is the frequency, Z is the length of the transmission line and C and R are the characteristic capacitances and resistances, respectively. It can be seen from
this equation that the amplitude of noise is an exponential function of the

length z. For a shielded lead $C_T$'s and $R_T$'s are constant. However, if the electrolyte and/or liquid ion exchanger filled shank of a typical microelectrode is exposed to air both $C_T$'s and $R_T$'s are variable. If the experiment is carried out in a Faraday cage the shield represents a fixed plate, the electromagnetic field around the microelectrode assembly is constant and noise interference is thus eliminated. Some experiments can be, undoubtedly, carried out under such conditions, however, if the experiment cannot be protected by a Faraday cage the best solution to the noise problem appears to be the reduction of the distance between the microelectrode and the preamplifier. The purpose of this communication was to show an arrangement which, we believe, has the shortest possible distance. It is pertinent to mention here that due to the small volume of the internal solution involved it is preferable to use a true ion buffer system rather than a solution of strong electrolyte such as KCl.

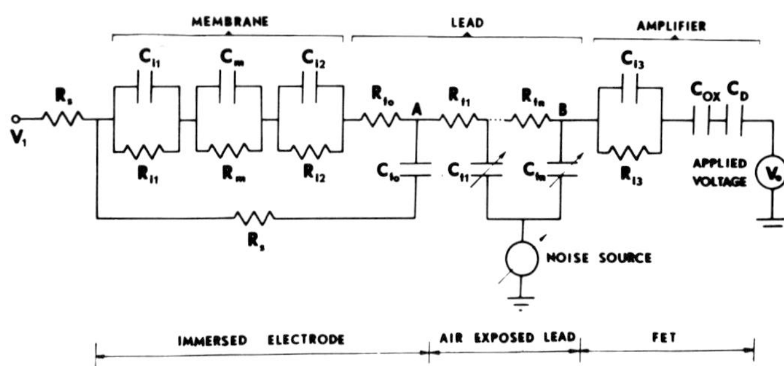


Fig. 4. Electrical equivalent circuit of an experiment involving a liquid ion exchange microelectrode, lead and a preamplifier

## References


1. Brown HM (1976) Intracellular $Na^+$, $K^+$, and $Cl^-$ activities in Balanus photoreceptors. J Gen Physiol 68:281
2. Fein H (1977) An introduction to microelectrode technique and instrumentation, 1st edition, W-P Instruments, New Haven, Connectivut, pp 1-23
3. Haemmerli A, Janata J, Brown HM (1980) Ion-selective electrode for intracellular potassium measurements. Anal Chem 52 (in print)
4. Janata J, Huber RJ (1979) Ion-sensitive field effect transistor. Ion-selective electrode review 1:31
5. Morf WE, Simon W (1978) Ion-selective electrodes based on neutral carriers. In: Freiser H (ed) Ion-selective electrodes in analytical chemistry, Vol 1, Plenum Press, New York, pp 211-286
6. Paris T, Hurd FK (1969) Basic electromagnetic theory. McGraw Hill, New York, p 425
7. Smith RL, Janata J, Huber RJ (1980) Transient phenomena in ion sensitive field effect transistors. J Electrochem Soc 127 (in print)
8. Thomas RC (1978) Ion-sensitive intracellular microelectrodes. Academic Press New York, pp 1-110

9.  Walker JL (1971) Ion specific liquid ion exchanger microelectrodes.
       Anal Chem 43:89A
10. Walker JL (1979) Ion selective electrode measurements. Methods in Enzy-
       mology 59:359
11. Wong B, Woody CD (1978) Recording intracellularly with potassium ion-
       sencitive electrodes from single cortical neurons in awake cats.
       Exp Neurol 61:219


Acknowledgements

This work was supported by NIGMS grant No. 22952.

Financial assistance for one of us (A.H.) was kindly provided by Ebauches
S.A. (Switzerland)


Department of Bioengineering, Department of Physiology, University of Utah,
Salt Lake City, Utah 84112/USA

# A Small Flexible pH-Electrode for Esophageal Monitoring

N.C. HEBERT, R.R. DELEAULT, A. FAUCON

A small flexible pH-electrode was developed for monitoring the esophageal pH of small infants (Fig. 1). The sensing part of the electrode consists of a 1.3 mm diameter pH glass hemisphere sealed onto a 1.3 mm glass tube. The total length of the glass electrode is 6 mm. A pH buffer with a Ag/AgCl internal reference electrode is completely glass sealed around a platinum pin and attached to the center core of a low-noised miniature coaxial cable. The outer sleeve of the coaxial cable is pushed out over the glass stem. A silicone rubber is used to insure a good, flexible water-tight seal around the glass.

The pH-electrode can be used with a standard laboratory reference electrode or with a skin reference electrode for in-vivo measurements.

## Specifications

| | |
|---|---|
| Total length | 3 m |
| Body (outer diameter) | 16 mm |
| Electrode (outer diameter) | 1.3 mm |
| pH range | pH 1 to 13 |
| Response time | 5 to 15 seconds |
| Depth of immersion | 1.0 mm |
| Noise level | Max. 0.03 pH |
| Temperature | $15^{\circ}$ to $50^{\circ}C$ |
| Rigidity | 10 mm long from tip |
| Electrical resistance | 1500 Megohms at $25^{\circ}C$ |
| Sterilization | Chemically |

## In-Vitro Calibration (Fig. 2)

The flexible pH-electrode along with a standard laboratory calomel or silver-silver chloride reference electrode with liquid-liquid junction can be used as the electrode pair. The electrode pair is calibrated in two buffer solutions. The first pH buffer is used to calibrate the zero offset of the electrode pair. The second pH buffer is used for slope adjustment.

## In-Vivo Calibration (Fig. 3)

A skin reference electrode is applied to the patient and the pH electrode is placed into a test tube containing a buffer solution. A pH reading is taken after the patient touches the buffer solution or a buffer soaked gauze extending from the buffer solution, (4). Several buffer solutions covering the pH range expected during an experiment are used, e.g. pH 1.4 and 7.

The pH-electrode can also be calibrated on the technician before measurements on a patient. This method of calibration can result in actual pH errors up to 0.5 pH units. Errors of this magnitude are not critical since the user is more interested in large pH changes rather than actual values.


## Sterilization

The electrode can be sterilized in ethylene oxide or disinfected in zephirin chloride solutions.


## Research Applications

Esophageal reflux (regurgitation of stomach contents into the esophagus) is a problem that affects many people. Left untreated serious complications including esophageal ulceration, esophageal stricture, and upper gastrointestinal bleeding can occur. Esophageal pH monitoring is a useful diagnostic tool for recording the number and severity of these reflux episodes in a given period. In addition, pH monitoring can indicate how well the esophagus is "cleared" by swallowing after an occurrence (1, 2, 3, 6).

Leape et al. (4) report incidents of respiratory arrest in infants secondary to gastroesophageal reflux. They have devised a system using a small flexible pH-electrode which is connected to a pH meter and recorder that monitors pH over many hours. Heart rate and respiration are also simultaneously monitored (5). Infants that have been resuscitated from an SIDS episode (Sudden Infant Death Syndrome) and show a relationship linking periods of respiratory arrest with periods of gastroesophageal reflux would strongly indicate surgical treatment.

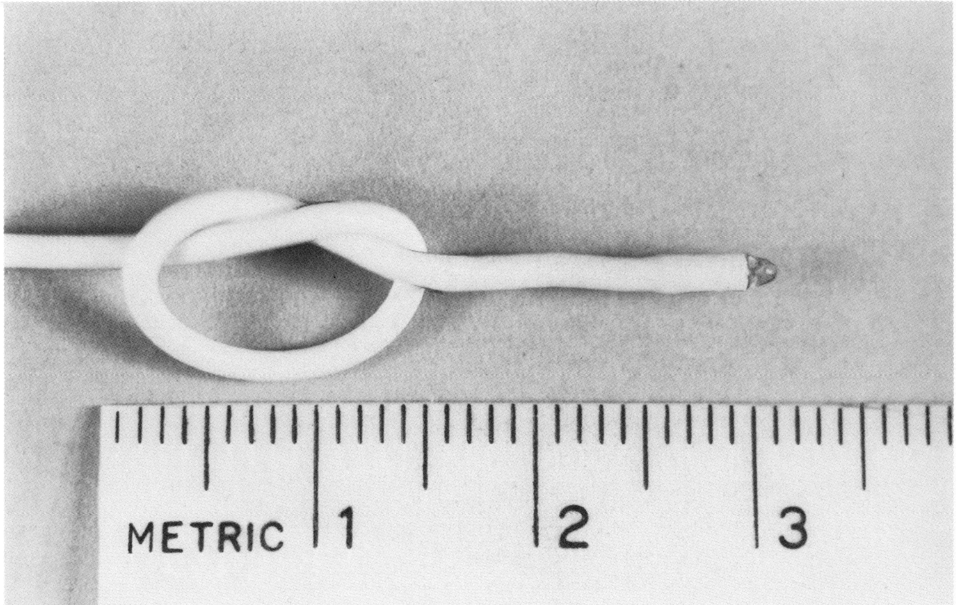


Fig. 1.  Photograph showing flexibility of small pH-electrode

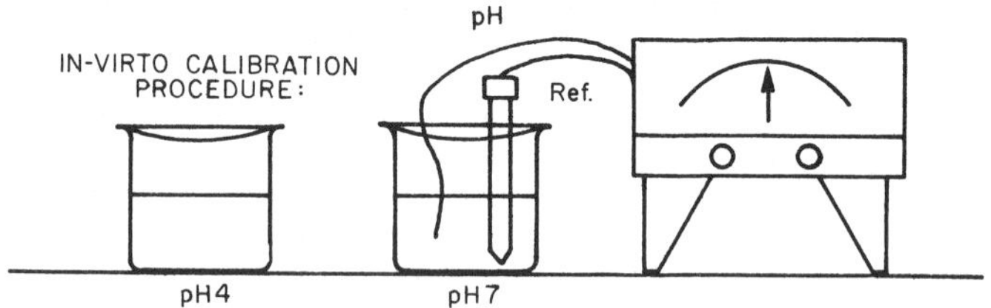



Fig. 2.   In-vitro calibration procedure

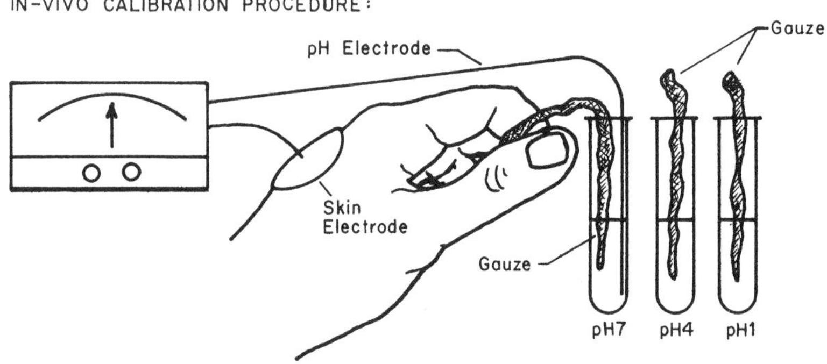



Fig. 3.   In-vivo calibration procedure

## References


1. Demeester TR, Johnson LF, Joseph GJ, Toscoro MS, Hall AW, Skinner DB
   (1976) Patterns of esophageal reflux in health and disease. An Surg
   184 (4): 459
2. Johnson LF, Demeester TR (1976) Medical therapy for gastroesophageal
   reflux assessed by 24-hour pH monitoring (Abstr). Gastroenteral 70:
   A39

3. Leape LL (1972) Gastroesophageal reflux as a cause of the Sudden Infant Death Syndrome. 78th Ross Conf Pediatr Res. Columbus, Ohio. Ross Laboratories
4. Leape LL, Holder TM (1977) Respiratory arrest in infants secondary to gastroesophageal reflux. Pediatrics 60: 924
5. Lee jr RJ, Connolly RJ, Romanofsky ML, Leape LL (1979) Integrated system for esophageal pH monitoring. 32nd Ann Conf Engin Med Biol October 1979. Alliance for Engineering in Medicine and Biology, Washington DC
6. Rosett NE, Dunscomb JE, Cuello LF, Rodriquez LJ, Cohen CM (1961) In situ pH gastric contents in patients with gastric and duodenal ulceration under therapy. Am J Gastroent 36: 553

Microelectrodes Inc., Grenier Industrial Village, Londonderry NH 03053/USA

# Bio-Electrodes for L-Histidine: A Pseudomonas Bacterial Electrode and a Histidine Ammonia-Lyase Enzyme Electrode

R.R. WALTERS*, R.P. BUCK

This investigation was aimed at comparing electrodes for detection and monitoring L-histidine. The bacterial electrode used pseudomonas sp. ATCC 11299b and purified histidine ammonia-lyase from the same bacterium provided the enzyme electrode. The enzymes for histidine degradation in pseudomonas were induced by the presence of histidine in the growth medium (3). Two moles of ammonia are produced from each mole of histidine (5). In our work temperature effects, pH effects and methods of immobilization have been examined thoroughly and two papers accepted by Analytical Chemistry Journal. A special contribution was an elevated temperature treatment that decreased the time needed to reduce high background levels of residual ammonia in the electrode.

The bacteria were obtained from a culture collection and grown at 25 – 30°C in medium of Tabor and Mehler (6). Growth was followed optically and bacteria harvested by centrifugation. The electrode preparation roughly followed Kobos and Rechnitz (2) using an ammonia gas sensing electrode; however we have found sterilizing filter, Millipore GS-0.22/um pore size, to be superior to dialysis membrane for containing the bacteria. Reaching low ammonia levels within the electrode, improving speed of response for baseline (zero histidine) values and superior mechanical strength and stiffness are advantages of the sterilizing filter.

The enzyme purification at 0° – 4°C followed methods of Rechler (4) and Givot et al. (1). High specific activity was achieved by gradient elution on DEAE cellulose after dialysis. Enzyme aliquots were lyophilized and stored at –10°C until used. Histidine ammonia-lyase, as an active ingredient in an electrode, presents difficulties. The enzyme contains four readily oxidized sulhydryl groups and two metal binding sites. Activity and kinetics depend on the oxidation state of the enzyme and the metal activator used. We have contributed a study of $Cd^{2+}$ and $Mn^{2+}$ as activators with complexing agents and glutathione pH effects, buffer components, and ionic strength were also investigated. Enzyme electrodes were made from ammonia gas sensors by placing up to 10/ul of enzyme solution between the gas permeable membrane and a circular dialysis membrane (Spectra/Por$^{-2}$, Fisher Scientific Co.) or by chemically immobilizing the enzymes with glutaraldehyde and bovine serum albumin. Electrodes were activated before use each day by soaking for 1 hr at 25°C in a 6.7 mM glutathione solution (pH approx. (.8)).

## Results – Bacterial Electrode

The linear range of a bacterial electrode is a portion of the mV-log activity curve between level responses at both high and low activities. This region is between $3 \times 10^{-5}$ and $3 \times 10^{-3}$ M histidine with a slope of 48 – 53 mV/decade. Increased high level response occurs at high (12 – 20 mg), wet-packed cells, while normal 10 – 15 mg gives a response becoming non-linear at 1 x $10^{-3}$ M histidine. However, heavy loading increases response times. Longest time responses (hours) occur at 25°C when background level (zero histidine) is sought. However, elevated temperature dialysis at 45°C for 15 minutes, then measurements at 25°C, produce extended regions of linearity, improved response times (7 - 12 min) and slopes and minimized dialysis times. On succeeding days an electrode required no further heat treatment. Baseline potentials were reached within 30 min each day after storage in buffer.

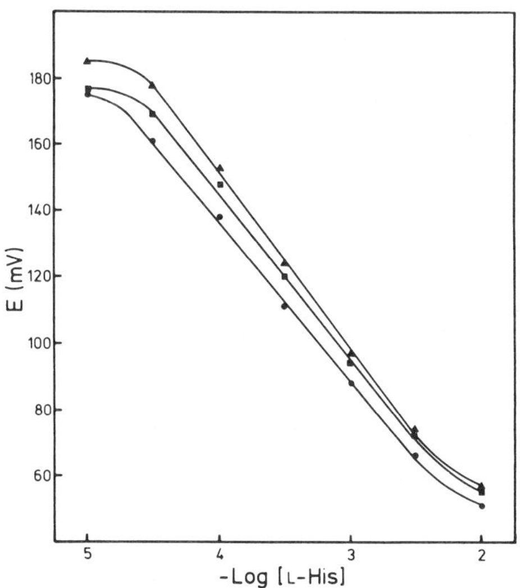


Fig. 1.   Response curves of a bacterial electrode containing 12 mg bacteria. (●) day 1, (■) day 9, (▲) day 22

The heat treatment did have two deleterious effects: the response curve at moderate histidine concentrations occasionally became super-Nernstian and low-level responses were lost. It is likely that after temperature lowering, the metabolic rate slowed and the bacteria no longer had enough ammonia for cell processes. When exposed to low histidine levels the ammonia produced may have been retained in the cells. Super-Nernstian responses suggest non-equilibrium production of ammonia. It is also likely that heat treatment killed some of the more sensitive bacteria.

The effect of pH on the bacterial electrode was examined at pH 7.5, 8.0 and 8.5. Although slopes are unaffected, the repsonse range was greatest at pH 8.5. Super-Nernstian slopes occurred at ionic strengths below 0.1. Major interferences were urocanic acid and urea. Arginine, tryptophane, glycine, lysine and glucose were not interferences. Electrode responses were studied up to 30 days. At 22 days, slopes actually improved (48 mV/decade to 53mV/decade). There is minor shift in the position of the response on the potential scale, so that calibrations are required. Electrodes stored in buffer give reproducibility of $\pm$ 1 mV. Of course, nitrogenous buffers are not desirable because of low-level interferences. Phosphate buffers are suitable.

<u>Results - Enzyme Electrodes</u>

Optimization of this enzyme electrode system is difficult. The purification procedure converted histidine ammonia-lyase to the oxidized form. Reduction of enzyme sulhydryl groups by glutathione, ß-mercaptoethanol or dethiothreitol converts the enzyme to its active form. Both $Mn^{2+}$ and $Cd^{2+}$ activate oxidized and reduced forms, although the reduced enzyme activity is markedly improved. We studied metal levels, pH (8.1 to 9.9) and complexing agents. Too high pH or too much complexing agents hurts activation by $Mn^{2+}$ and $Cd^{2+}$, presumably by extensive hydrolysis and complexation. Some buffer components are inhibitors; we settled on 0.1 M Tris buffer, pH 9.7, $10^{-5}$ M $Mn^{2+}$, 2 x $10^{-5}$ M citrate and 3 x $10^{-4}$ M glutathione. Then the electrode response is linear from 3 x $10^{-5}$ M to $10^{-2}$ M histidine with a slope of 54 mV/decade. The response times to 3 - 8 minutes, 0.08 U to 0.8 U (U = units) of enzyme per electrode is adequate. Higher loading eliminates inhibitor effects and slightly improves response times, slopes and reproducibility.

Immobilization of the enzyme beneath a dialysis membrane used bovine serum albumin-glutaraldehyde. Time responses increased to 8 - 12 min although responses were otherwise the same, Fig. 2. Ordinary trapped enzyme electrodes were preferred.

Responses were not improved significantly at elevated temperatures. Electrodes calibrated every eight days reproduced to $\pm$ 2 mV from 3 x $10^{-5}$ M to $10^{-2}$ M. Lifetimes were at least 1 month. Possible interferences tested for the bacterial electrodes were not interferences for the enzyme electrodes. Ammonia is the principle interference. At equal concentrations with histidine, glycine and urocanic acid are slight inhibitors (few millivolt effect).


Comparison of Bacterial and Enzyme Electrodes for Histidine.

Table 1 summarizes some of the properties. Both have similar sensitivity to histidine, but the enzyme electrode is clearly superior in terms of most analytical properties. The purified enzyme required about one week to prepare, but enough enzyme was obtained for 70 electrodes, and the one enzyme supply considering lifetime, should give sufficient probes for several years. The bacterial electrode is more durable. It is not harmed by pH or temperature changes relative to the enzyme electrode.

Interferences are probably the decisive factors in choosing between electrode formulations. The bacterial electrode is not suitable for serum histidine because of high urea (equivalent to $10^{-3}$ M histidine) interference.

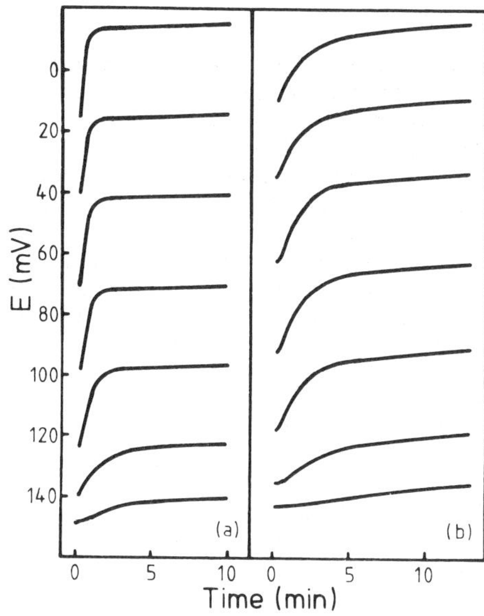

Fig. 2. Time responses of electrodes containing enzyme immobilized by a dialysis membrane (a) and by glutaraldehyde crosslinking with BSA (b). The histidine concentrations are, from top to bottom, $10^{-2}$ M, $3 \times 10^{-3}$ M, $10^{-3}$ M, $3 \times 10^{-4}$ M, $10^{-4}$ M, $3 \times 10^{-5}$ M, $10^{-5}$ M

Table 1. Comparison of histidine enzyme and bacterial electrode characteristics

| | | |
|---|---|---|
| Response slope | 52–54 mV/decade | 47–56 mV/decade |
| Linear range | $3 \times 10^{-5}$ M to $10^{-2}$ M | $3 \times 10^{-5}$ M to $3 \times 10^{-3}$ |
| Response time | 3 – 8 min | 7 – 12 min |
| Lifetime | More than 8 days | 3 weeks |
| Interferences | Ammonia + a few minor negative interferences | Ammonia + several moderate positive interferences |
| Day-to-day resproducibility | Excellent | Good |
| Initial electrode preparation | Thaw enzyme and assemble | Grow and harvest bacteria and assemble |
| Requirements for operation | Requires daily activation and the addition of a thiol $n^{2+}$ to each sample | Requires only an initial heat treatment to reduce ammonia levels |

## References

1. Givot IL, Smith TA, Abeles RH (1969) Studies on the mechanism of action and the structure of the electrophilic center of histidine ammonia lyase. J Biol Chem 244: 6341
2. Kobos RK, Rechnitz GA (1977) Regenerable bacterial membrane electrode for L-asparate. Anal Lett 10: 751
3. Lessie TG, Neidhardt FC (1967) Formation and operation of the histidine-degrading pathway in pseudomonas aeruginosa. J Bacteriol 93: 1800
4. Rechler MM (1969) The purification and characterization of L-histidine ammonia-lyase (pseudomonas). J Biol Chem 244: 551
5. Tabor H, Hayaishi O (1952) The enzymatic conversion of histidine to glutamic acid. J Biol Chem 194: 171
6. Tabor H, Mehler AH (1955) Histidase and urocanase. From histidine ammonia-lyase enzyme electrode for determination of L-histidine. Methods Enzymol 2: 228

This work was supported by National Science Foundation, Grant CHE77-20491.

Department of Chemistry, University of North Carolina, Chapel Hill, No. Car. 27514 USA

*Present address: Department of Chemistry, Iowa State University, Ames, Iowa, USA

# Recent Developments in the Field of Enzyme Membrane Electrodes

Z. STEFANAC[*], W.E. MORF, U. THANEI, I. MOSTERT, R. DÖRIG, H.B. JENNY, and
W. SIMON

## Introduction

Potentiometric enzyme electrodes have been designed for the determination
of several clinically relevant species (1-3). They typically consist of an
enzyme layer covering the surface of an ion-selective electrode or a gas
sensor which measures the products released by enzymatic degradation of
substrates. More recently, continuous-flow systems were introduced in which
separate enzyme reactors are coupled on-line to the appropriate sensor
units (2, 4, 5). The salient feature is that the reaction and the detection
step can be controlled and optimized independently.

## Theoretical Considerations

Recently, a simple theoretical model was evolved that provides a direct
comparison between the performance of dipping-type enzyme electrodes and
that of flow-through enzyme reactors (6, 7). The following approximation
was obtained for the steady-state response of enzyme electrodes:

$$\text{EMF} = E^O + s \cdot \log \left[ K_{PS}^{Pot} [S] + Q \right] \tag{1}$$

where $E^O$ is a constant reference potential, $s$ is the slope of the response function. $[S]$ is the concentration of substrates in the sample solution,
and $Q$ is an interference term including any contributions from other
species. The selectivity factor $K_{PS}^{Pot}$ represents a measure of the sensitivity of the electrode toward substrates, being determined as:

$$K_{PS}^{Pot} = \frac{\bar{n} \, k_R}{k_R + k_S + k_S \, \dfrac{K_S[S]}{K_M}} \tag{2}$$

Here $\bar{n}$ is the maximum number of measurable products derived per substrate molecule, $k_R$ and $k_S$ are kinetic parameters ($s^{-1}$) characterizing
the rate of the over-all enzymatic reaction and that of the mass transfer,
$K_S$ is the distribution coefficient for substrates between sample solution
and enzyme membrane, and $K_M$ is the true Michaelis constant (mol/l) (7). According to Eqs. (1) and (2), enzyme electrodes are capable of measuring
substrates only in a limited concentration range where, ideally, a Nernstian response is observed:

$$\text{EMF} = \text{const} + s \cdot \log [S] \tag{3}$$

The detection limits are found at the following concentrations:

62

$$[S]_{min} = Q/K_{PS}^{Pot} \tag{4}$$

$$[S]_{max} = K_M^{app} = \frac{K_M}{K_S} \left(1 + \frac{k_R}{k_S}\right) \tag{5}$$

where $K_M^{app}$ is the apparent Michaelis constant characteristic of an enzyme immobilized in a membrane. Below and above these limits the electrode response becomes invariant with the substrate concentration. Since the pivotal parameter $k_R$ depends on the concentration $[\bar{E}]$ of immobilized active sites and on the turnover number $k_2$ of the enzymatic reaction (7),

$$k_R = k_2 [\bar{E}] / K_M, \tag{6}$$

a high enzyme activity must evidently be provided in electrodes in order to extend their analytically useful range. Otherwise, the response characteristics remain extremely sensitive to changes in the enzyme activity, a fact which is often observed experimentally (3, 6).

In enzyme reactors the flowing solution is exposed to the fixed enzyme layer for a sufficiently long time in order that a large fraction $\eta$ of substrate be converted to product (s). The following implicit relation was derived for the conversion efficiency $\eta$ (5, 7):

$$[S]_o \, \eta - K_M^{app} \ln(1-\eta) = v_{max} \, t \tag{7}$$

$[S]_o$ is the substrate concentration of the sample solution before it enters the reactor, $K_M^{app}$ is the constant introduced in eq. (5), and $v_{max}$ is the maximal conversion rate of the reactor, being defined as:

$$v_{max} = \frac{\bar{V}}{V} k_R K_M = k_2 \frac{\bar{V}[\bar{E}]}{V} \tag{8}$$

where $V$ and $\bar{V}$ are the volumes of the mobile phase and of the active enzyme layer, respectively. If a potentiometric sensor is coupled on-line to the outlet of the enzyme reactor, the concentration of the n product species formed from each reacting substrate molecule gives rise to the following EMF-response:

$$EMF = E^{O\prime} + s \cdot \log \left[n \eta \cdot [S]_o + Q\right] \tag{9}$$

Calculations based on eqs. (7) and (9) show that the response pattern of enzyme reactors is quite similar to that found for the conventional enzyme electrodes, as long as $v_{max} t < K_M^{app}$. In contrast, significant improvements can be realized for $v_{max} t \gg K_M^{app}$ where a conversion efficiency of $\eta \cong 1.0$ results even for high substrate concentrations. Accordingly, a linear response of the type (3) is exhibited by corresponding enzyme-reactor-electrode cells, the detection limits being given by

$$[S]_{o,min} = Q/n \tag{10}$$

$$[S]_{o,max} = v_{max} \, t \gg K_M^{app} \tag{11}$$

A relatively high enzyme activity and a sufficiently long reaction time are found to be essential for high-performance enzyme reactors.

Previously described reactor-electrode combinations (2, 4, 5) showed promising characteristics. However, these continuous-flow systems are not optimally adapted to small sample volumes because here the reaction time (i.e. the time required for the mobile phase to traverse the reactor) is fixed by the dimensions of the reactor. Hence a discontinuous technique must be chosen for a microdetermination of substrates. A new, miniaturized enzyme reactor cell of this type is described in the following.


An Enzyme-Reactor-Electrode Cell Suited for Microdeterminations of Urea

A schematic diagram of the miniaturized cell is shown in Figure 1. The reactor (plexiglass tube, length $\sim$ 3 cm, inner diameter O.3 cm) was packed with porous glass beads on which the enzyme urease was immobilized (4, 8). The outlet of the reactor projects into the measuring chamber (plexiglass cell, minimal volume $\sim$ 100 $\mu$l) which was equipped with an indicator and a reference electrode. An ion-selective minielectrode based on nonactin/monactin (8) was used for the measurement of ammonium ions released by the enzymatic hydrolysis of urea, and a calomel electrode in contact with a 1M lithium acetate salt bridge was employed as reference half-cell. Measurements were performed at 25$^{\circ}$C.

A discontinuous technique was devised for the enzymatic analysis of very small substrate samples. After preconditioning the reactor for a short time with buffer solution (O.1M Tris-HCl, pH 7.0), the 10-$\mu$l sample was injected into the reactor inlet. The sample was passed through the reactor by applying vacuum to the outlet. The sample did not come into contact with the ion-selective electrode, which is believed to reduce interference problems. The reactor was then filled with 100$\mu$l buffer solution, and the reaction in the enzyme layer was allowed to continue for about three minutes. Afterwards, the solution was expelled with compressed air into the electrode compartment where the EMF-response was measured. The washing-out procedure was repeated and a second EMF-reading was taken for control purposes.

Figure 2 shows the calibration plot of the reactor-electrode cell for aqueous urea standards. A linear response was found for the studied range of $5 \cdot 10^{-4}$ to $10^{-1}$M urea with a slope of 61.2+1.1mV/decade. After eight days of continuous use, the reactor still contained a surprisingly high activity of urease although the upper detection limit was somewhat lowered and an E$^{\circ}$-shift of around 7mV was observed (see open circles in Figure 2).

In order to demonstrate the capability of the new enzymatic method it was applied to the analysis of blood serum. Serum samples were obtained from a clinical laboratory where urea concentrations are determined routinely using a colorimetric method after reaction with biacetyl mono-oxime. In our experiments, a 10-$\mu$l aliquot was taken from each sample and was injected into the reactor without further treatment. The results are given in Figure 3. The correlation on the left documents that the urea concentrations determined from EMF-readings in the first washing-out solutions compare favorably with the values found from the standard laboratory method. More surprisingly, the values obtained from control measurements in the second washing-out solutions also agree with routine measurements, as is shown on the right-hand side in Figure 3.

64

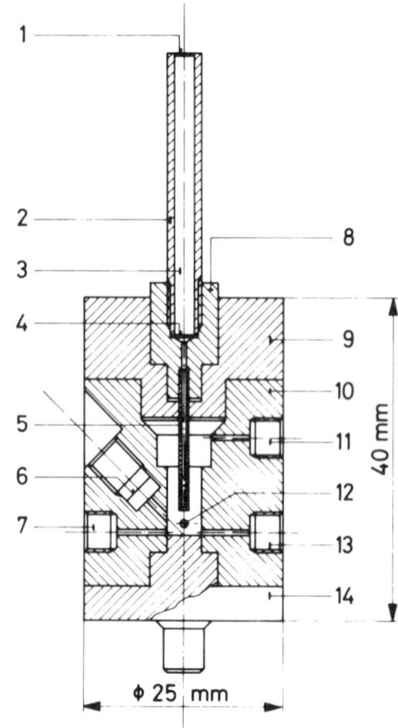

Fig. 1. Schematic diagram of the enzyme-reactor-electrode assembly. 1 -
reactor inlet (platinum frit), 2 - plexiglass tube, 3 - urease immobilized
on glass, 4 - nylon netting, 5 - reactor outlet, 6 - ammonium-selective
minielectrode, 7- to waste, 8, 9 - reactor support (plexiglass), 10, 14 -
measuring block (plexiglass), 11 - vent, 12 - liquid junction to reference
half-cell (diaphragm), 13 - inlet for rinsing solution

Fig. 2.   EMF-response of the cell to aqueous urea samples (10 $\mu$l aliquots)

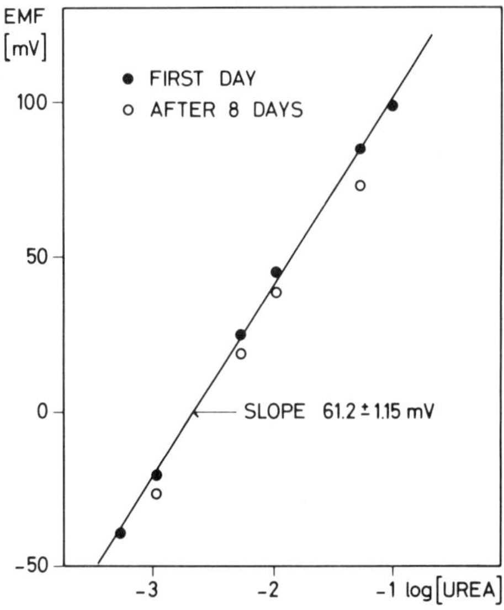

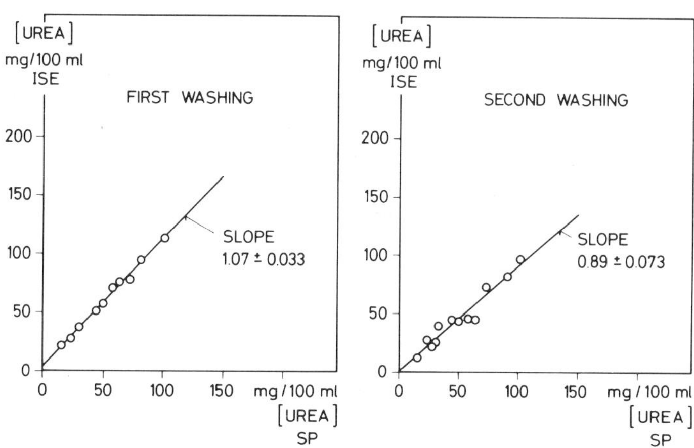



Fig. 3. Results of urea determinations in serum samples. Urea concentrations determined from EMF-measurements in the first and the second washing-out solutions are correlated with the values obtained from the biacetyl mono-oxime method

The present preliminary results suggest that, in principle, enzyme-reactor-electrode cells can be adapted to microdeterminations of substrates. It appears that further improvement is possible through minimal modification of the method.


## References


1. Buck RP (1978) Ion-selective electrodes. Anal Chem 50: 17R
2. Gray DN, Keyes MH, Watson B (1977) Immobilized enzymes in analytical chemistry. Anal Chem 49: 1067A
3. Guilbaut GG, Montalvo JG (1970) An enzyme electrode for the substrate urea. J Amer Chem Soc 92: 2533
4. Johansson G, Ögren L (1976) An enzyme reactor electrode for urea determinations. Anal Chim Acta 84: 23
5. Johansson G, Ögren L (1977) Application of ion-selective electrodes to enzymatic analysis. In: Pungor E (ed) Ion-selective electrodes. Akadémiai Kiadó, Budapest, p 93
6. Morf WE (1980) The principles of ion-selective electrodes and of membrane transport. Akadémiai Kiadó, Budapest
7. Morf WE (to be published)
8. Stefanac Z, Thanei U, Mostert I, Dörig R, Jenny H-B, Morf WE, Simon W (to be published)

Acknowledgement

The present work was partly supported by the Swiss National Science Foundation and the Eidgenössische Stiftung zur Förderung der schweizerischen Volkswirtschaft durch wissenschaftliche Forschung.

One of us (Z.St.) gratefully acknowledges financial support by the Scientific Exchange agreement.

We also thank Dr. R. Asper (Medico-chemical Central Laboratory, University Hospital, Zürich, Switzerland) for the supply of fresh serum samples as well as for the corresponding urea determinations.

Department of Organic Chemistry, Swiss Federal Institute of Technology, CH-8092 Zürich, Switzerland
* Present address: Zavod za Analiticku Kemiju, Prirodoslovno-Matematicki Fakultet, Sveuciliste u Zagrebu, Strossmayerov Trg 14, YU-41000 Zagreb, Yugoslavia

# Lactate Measurements with an Enzyme Optode that Uses Two Oxygen Fluorescence Indicators to Measure the PO$_2$ Gradient Directly

D.W. LÜBBERS, K.-P. VÖLKL, U. GROSSMANN, N. OPITZ

The main topic of this symposium is the measurement of ions such as $K^+$, $Na^+$, $Ca^{2+}$ etc. by means of potential differences produced by specific ion carriers across a fluid layer or a suitable membrane. I would like to mention that absorption as well as fluorescence measurements with specific indicators represent additional approaches to the measurement of these ions. The difficulty with such measurements is that the indicators often react with other substances in the sample whereby the calibration curve is changed. This restricts their applicability in biology. Our experiments have shown that this difficulty can be overcome by introducing a specialized membrane between probe and indicators which allows only the passage of those molecules which do not disturb the optical signals. The compartmentation by the membrane has several advantages: 1) the indicator can be dissolved in liquids which could disturb the sample, 2) the membrane can be used for an optical separation from the probe and 3) in some cases the indicator can be directly bound to the membrane. Fluorescence indicators are especially effective (3, 4), because of their high sensitivity.

In my report I will discuss another method of ion measurement. This method takes advantage of the fact that for some ions, as for example lactate, specific enzymes exist which change this substance into products which can be continuously monitored. So, the lactate oxydase E.C.1.13.12.4 (LOD) converts in the presence of oxygen lactate to acetate in the following way:

$$\text{lactate} + O_2 \xrightarrow{\text{LOD}} \text{acetate} + CO_2 + H_2O$$

Analogous to the enzyme electrode according to Updike and Hicks (7) and Clark (2), in which the $Po_2$ change caused by the oxidation of a substrate is measured by a platinum electrode, we applied fluorescence photometry to measure the oxygen pressure by a device which we call an "optode" (4). The advantage of the fluorescence measurement of oxygen is that in a steady state no oxygen is consumed by the indicator. But with the enzyme optode there is another more important advantage: since the enzyme layer is transparent one is able to use two different fluorescence indicators with different spectral characteristics simultaneously, one in front (indicator 1), the other on the backside of the enzyme layer (indicator 2) to measure directly the $Po_2$ gradient across the enzyme layer. Since with such a device the substrate concentration can be directly measured, it can be applied to local measurements in tissue.

For the fluorescence photometric measurement of $Po_2$ Knopp and Longmuir (3) proposed pyrenebutyric acid as the indicator. The fluorescence signal of pyrenebutyric acid is quenched by oxygen; with increasing $Po_2$ the intensity of the fluorescence signal decreases. According to Vaughan and Weber (8) the following equation is valid for low to medium oxygen concentrations corresponding to an oxygen pressure up to a $Po_2$ of ca. 400 mmHg (53.2 kPa):

$$\frac{I_O}{I} = 1 + K \cdot [O_2] \tag{1}$$

with $I_O$ as the fluorescence intensity in the absence of oxygen; $I$, fluorescence intensity of the respective oxygen concentration $[O_2]$; $K$, quenching constant.

The oxygen concentration is proportional to $Po_2$; the proportionality factor is the solubility coefficient, $a$:

$$[O_2] = a \cdot Po_2 \tag{2}$$

By substitution of this equation in eq. (1) we find

$$\frac{I_O}{I} = 1 + K \cdot a \cdot Po_2 \tag{3}$$

or

$$Po_2 = \frac{I_O - I}{a \cdot K \cdot I} = \frac{1}{a \cdot K} \cdot \frac{\Delta I}{I} \tag{4}$$

with

$$\Delta I = I_O - I$$

Eqs. (3) and (4) describe the fluorescence behaviour of the indicator layer with changing oxygen pressure.

Since the substrate (e.g. lactate) is oxidized in the enzyme layer, substrate and oxygen are transported along the concentration and pressure gradient into the enzyme layer. Applying the laws of diffusion for the steady state (see apendix), it is found that the substrate concentration in the sample, $[s]$, is proportional to the $Po_2$ difference between the $Po_2$ of the sample, $p(O)$ and that at the backside of the enzyme layer, $p(d)$.

$$[s] = c(o) = K_S^{-1} \cdot (p(o) - p(d)) = K_S^{-1} \cdot \Delta Po_2 \tag{5}$$

for information on $K_S$ and $c(o)$, see appendix

or

$$\Delta Po_2 = K_S \cdot [s] \tag{6}$$

Using two different fluorescence indicators to measure $p(o)$ and $p(d)$, respectively, we obtain with equation (4) and (6)

$$\frac{I_{o1} - I_1}{(a \cdot K)_1 \cdot I_1} - \frac{I_{o2} - I_2}{(a \cdot K)_2 \cdot I_2} = K_S \cdot [s] \tag{7}$$

or

$$K_1 \frac{\Delta I_1}{I_1} - K_2 \frac{\Delta I_2}{I_2} = K_S \cdot [s] \tag{8}$$

where $K_1 = (a \cdot K)_1^{-1}$; and $K_2 = (a \cdot K)_2^{-1}$

When p(o) is known the following equation is obtained

$$\frac{I_{o2}}{I_2} = 1 + (a \cdot K)_2 \cdot p(o) - (a \cdot K)_2 \cdot K_S \cdot [S] \qquad (9)$$

For the practical construction of the lactate optode it is important that the indicator layer (1) which is in contact with the probe does not inhibit diffusion of oxygen or substrate. It is therefore generally of advantage if the diffusion resistance of the indicator layer is small as compared with that of the enzyme layer. Since for the fluorescence photometric $Po_2$ measurements we always used condensed aromatic hydrocarbons dissolved in hydrophobic solutions, such a layer would be an insuperable barrier to the diffusion of substances such as lactate. For this reason the indicator (1) was encapsulated into nanocapsules (4) and then embedded in agar gel. As the indicator (1) pyrenebutyric acid was used. It was excited at $\lambda = 342$ nm and the emitted fluorescence was measured at $\lambda = 395$ nm. The second indicator perylene was found to be suitable since it can be excited at $\lambda = 436$ nm and emits its fluorescence at $\lambda = 490$ nm. The perylene solution was embedded in Di(2-ethylhexyl)phthalate. The viscosity of the layer was increased by Ethocel (ethylcellulose). Fig. 1 shows the excitation spectra of pyrenebutyric acid and perylene.

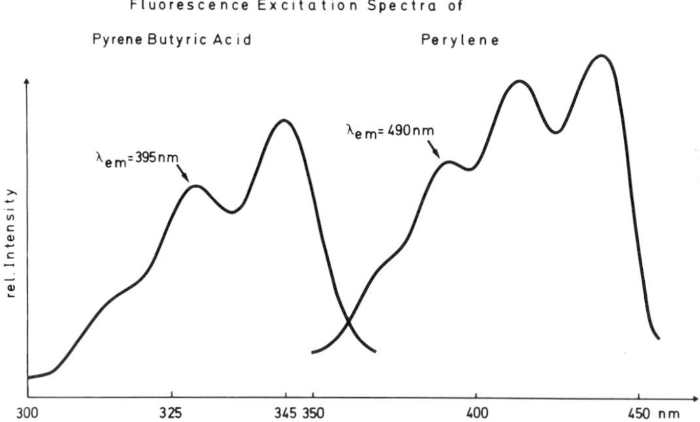


Fig. 1. Fluorescence excitation spectra of pyrenebutyric acid and perylene (relative intensity vs. wavelength). The emission is measured 1) with pyrenebutyric acid at $\lambda_{em} = 395$ nm; and 2) with perylene at $\lambda_{em} = 490$ nm

Fig. 2 shows a schematic drawing of the enzyme optode. In the cutout the
different layers are to be seen; the enzyme layer is covered by a thin cel-
lophane membrane with the gel which contains the nanoencapsulated indicator
(1) (C, $In_1$). The layer with indicator (2) is situated behind the enzyme
layer and is protected with a teflon membrane (T,$In_2$). The total layer
sandwich lies on a UV permeable plexiglass (P) through which excitation and
fluorescence light passes. The measuring chamber lies in front of the mul-
tilayer assembly. The fluorescence of both indicators was continuously monitored by the filter wheel photometer of Chance et al. (1). Using a flexible light guide (L) (diameter 4 mm), the enzyme optode can be directly
brought into contact with organ surfaces after removal of the chamber cup
(volume $\approx$ 100/ul). The enzyme optode allows continuous determinations in
biological fluids without interfering with the biological system: Only a
small part of the substance to be investigated diffuses through the cello-
phane membrane covering the enzyme optode and is then oxidized in the enzyme layer; larger molecules are retained. It allows the measurement of
both $Po_2$ and concentration of the substrate at the same site. The further
advantage of the optical method is that neither mechanical contact to the
site of measurement nor electrical connections are necessary. The optode
can be miniaturised and is relatively atraumatic. Fig. 3 shows a calibration curve for lactate according to equation (9).

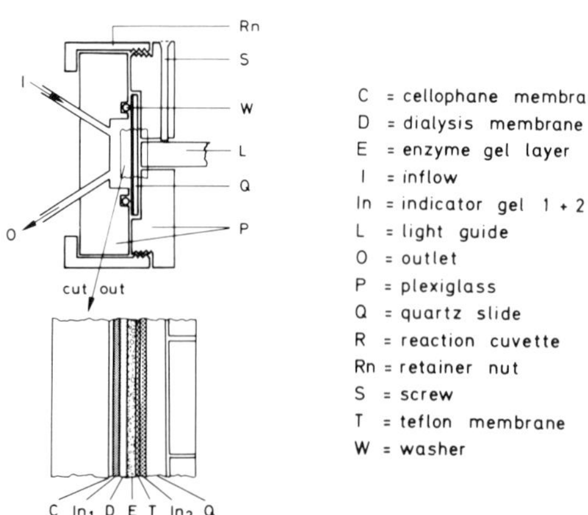



Fig. 2.  Schematic drawing of the lactate optode, (for details see text).
The rectangular field around the light guide is enlarged in the cutout,
the black line represents the membrane assembly

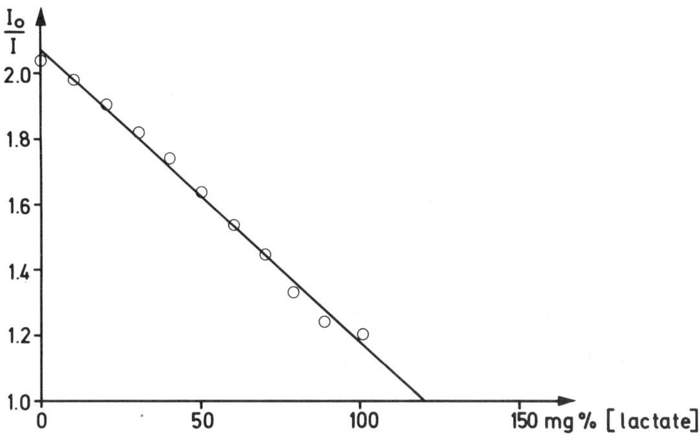

Fig. 3:  Calibration curve of the lactate optode, (fluorescence ratios vs. lactate concentration)


Appendix: Mathematical Model

We assume
1) that in a layer of a thickness, d the substrate, S is oxidized by a homogeneously distributed enzyme;
2) that the frontside of the enzyme layer is in contact with the probe, that at this boundary layer the concentration of the substrate and the pressure of oxygen corresponds to the values in the probe and that these values are constant;
3) that the backside of the enzyme layer is "impermeable" to substrate and oxygen;
4) that the profiles of substrate concentration and oxygen pressure within the enzyme layer, which are produced by the oxidation of the substrate can be calculated according to the laws of diffusion.

To describe the decrease of oxygen pressure, p(x) and substrate concentration, c(x) from the boundary probe/enzyme layer (x = 0) to the backside of the enzyme layer (x = d) we have to solve the following boundary value problems:


$$O = p''(x) - A_1 \cdot c(x) \quad ; \quad 0 \le x \le d \tag{10}$$

$$O = c''(x) - A_2 \cdot c(x) \quad ; \quad 0 \le x \le d \tag{11}$$

$$c(O) = \text{constant} \tag{12}$$

$$p(O) = \text{constant} \tag{13}$$

$$p'(d) = O = c'(d) \tag{14}$$

72

where

$$A_1 = \frac{E \cdot (V_m)_n}{F \cdot d \cdot c^* \cdot D_1 \cdot \dot{c}} \qquad ; \qquad A_2 = \frac{E \cdot M_r}{F \cdot d \cdot c^* \cdot D_2} = \frac{\dot{c}}{c^* \cdot D_2} = s^2$$

| | |
|---|---|
| $E$, | enzyme activity ($1.25\,\mu mol \cdot min^{-1}$); |
| $(V_M)_n$, | normal mol volume ($22.400\ l \cdot mol^{-1}$); |
| $M_r$, | molecular weight ($89\ g \cdot mol^{-1}$); |
| $D_1$, | diffusion coefficient for oxygen ($2 \cdot 10^{-5}\ cm^2 \cdot s^{-1}$); |
| $a$, | oxygen solubility ($2.2\ ml\ O_2/(100\ g \cdot 100\ kPa)$); |
| $F$, | area of the enzyme layer ($1\ cm^2$); |
| $d$, | thickness of the enzyme layer ($2.5 \cdot 10^{-3}\ cm$); |
| $D_2$, | diffusion coefficient of lactate, ($1 \cdot 10^{-6}\ cm^2 \cdot s^{-1}$); |
| $c^*$, | a defined substrate concentration ($c^* = 1400\ mg\%$) (see below); |

$c^*$ is the substrate concentration which under standard conditions would cause a substrate consumption of

$$\dot{c} = E \cdot \frac{M_r}{F \cdot d}$$

in the enzyme layer.

We obtain the substrate concentration c by solving (11) together with (12) and (14):

$$c(x) = c(o)\ \frac{\cosh\,(s \cdot (d-x))}{\cosh\,(s \cdot d)} \qquad ; \qquad o \leq x \leq d \qquad (15)$$

So substituting (15) into eq. (10) and then solving (10) together with (13) and (14) yields the oxygen partial pressure

$$p(x) = p(o) - \frac{A_1}{A_2} \cdot ((c(o) - c(x)) \qquad ; \qquad o \leq x \leq d \qquad (16)$$

With eq. (15) and (16) the dependence of the substrate concentration in the medium, $c(o)$ on the $P_{O_2}$ decrease across the enzyme layer ($p(o)-p(d)$) is found to be

$$c(o) = \frac{A_2}{A_1} \cdot \frac{1}{1 - \dfrac{1}{\cosh\,(s \cdot d)}} \cdot (p(o) - p(d)) \qquad (17)$$

or

$$c(o) = S = K_s^{-1} \cdot (p(o) - p(d)) \qquad (5)$$

Fig. 4 shows the calculated decrease of oxygen pressure and lactate concentration across the enzyme layer. The calculated $P_{O_2}$ agrees well with the measured one.

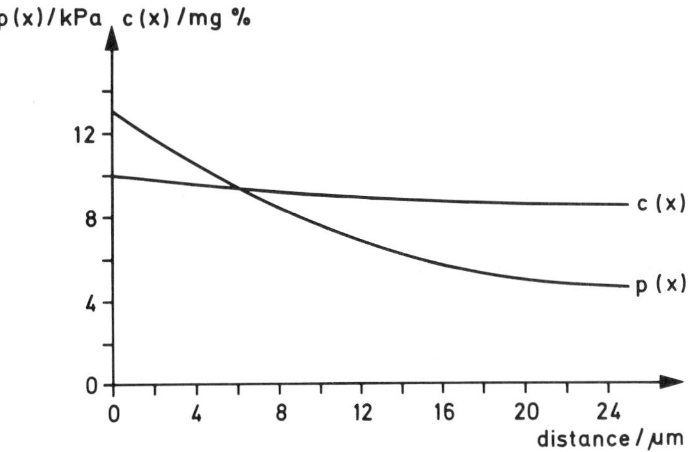


Fig. 4. Calculated lactate concentration and oxygen pressure profiles in the enzyme layer, (p(x) or c(x) vs. distance across the membrane oxygen pressure, p(x) in kPa, lactate concentration, c(x) in mg%)

## References


1. Chance B, Legallais V, Sorge J, Graham N (1975) A versatile time-sharing multichannel spectrophotometer, reflectometer, and fluorometer. Analyt Biochem 66: 498

2. Clark LC jr, Champ L (1968) Electrode systems for continuous monitoring in cardiovascular surgery. Am NY Acad Sci 102: 29

3. Knopp JA, Longmuir JS (1972) Intracellular measurements of oxygen by quenching of fluorescence of pyrene butyric acid. Biochem Biophys Acta 279: 393

4. Lübbers DW, Opitz N (1975) Die $pCO_2/Po_2$-Optode: Eine neue $pCO_2$- bzw. $Po_2$-Meßsonde zur Messung des $pCO_2$ oder $Po_2$ von Gasen und Flüssigkeiten. Z Naturforsch 30c: 532

5. Lübbers DW, Opitz N, Speiser PP, Bisson HJ (1977) Nanoencapsulated fluorescence indicator molecules measuring pH and $Po_2$ down to submicroscopical regions on the basis of the optode principle. Z Naturforsch 32c: 133

6. Opitz N, Lübbers DW (1975) A new fast-responding optical method to measure $Po_2$ in gases and solutions. Pfluegers Arch 355: R120

7. Updike SJ, Hicks GP (1967) The enzyme electrode. Nature 214: 396

8. Vaughan WM, Weber G (1970) Oxygen quenching of pyrene butyric acid fluorescence in water. A dynamic probe of the microenvironment. Biochemistry 9: 464



Max-Planck-Institut für Systemphysiologie, Rheinlanddamm 201, 4600 Dortmund 1/FRG

# Calibration Standards for Multi Ion Analysis in Whole Blood Samples

H.F. OSSWALD, H.R. WUHRMANN

Several analytical devices on the market today are capable of determining various parameters in whole blood with the aid of ion-selective electrodes. These devices are generally suitable for use in intensive care units or in the field of anaesthetics. The results of a measurement can be made available only a few minutes after the sample is taken. The necessary electrode calibrations are generally carried out using a minimum of two standard solutions and the results are used to determine the calibration lines. It has become common practice to perform a one point calibration in addition to every measurement of a sample in order to correct any possible shifts of the electrode potential. For practical and economic reasons, it is customary to use as calibration standards two aqueous solutions having electrolyte concentrations lying approximately within the physiological range of blood serum. Up to the present it has been the practice in clinical chemistry to give the results of electrolyte determinations in the form of concentrations, with the exception of the pH value. However, when a measurement is made with the aid of ion-selective electrodes, it is the activity which is being measured. Errors in the concentration determinations are to be expected:
1) if the ionic strength of the standard solution does not conform to that of the substance being measured, and
2) if there are differences in the liquid junction potential between standard solutions and the sample being measured.

This paper discusses a number of theoretical and practical aspects which should be regarded when planning a measurement system based on two standard calibration solutions. These standards are intended to enable the simultaneous calibration of four different ion-selective electrodes for measurements in whole blood samples. The device which is considered as an example is designed to determine the concentration of $K^+$, $Ca^{2+}$ and $Na^+$ ions simultaneously as well as the pH value. The following electrodes are used for this purpose: Capillary-glass membrane electrodes for the detection of $Na^+$ and $H^+$ ions ($pNa_{1000}$-glass or T- or Lot-glass) and neutral carrier-based liquid membrane electrodes for the determinations of $K^+$ and $Ca^{2+}$ ions, such as those described in the contribution of Dr. D. Ammann. The choice of these electrodes ensures that interfering ions present in the blood do not affect the results of the measurements by more than 1%. Also these electrodes showed theoretical slopes over a longer time period (typically over several weeks), both at room temperature and at $37^{\circ}C$, when whole blood and plasma samples are used. Since the human body balances out deviations in activity, the range of normal values for the electrolytes is very narrow. Since the signal of the ion-selective electrode is logarithmically dependent on the activities of the ions in the substance being measured, large differences in concentration do not necessarily have to result in large changes in the signal. Column 4 of Table 1 shows the required accuracy of analysis called for in medicine today and column 5 shows the corresponding accuracy of the appropriate electrode.

Table 1. Normal ranges and required accuracies of electrolytes in clinical chemistry

| ion | normal range | $\Delta$EMF [mV] | required accuracy ($\pm$ 1 SD) | EMF [mV] |
|---|---|---|---|---|
| $H^+$ | 7.35 - 7.45 | 5.9 | $\pm$ 0.01 | $\pm$ 0.59 |
| $Na^+$ | 135 - 145 mmol/l | 1.8 | $\pm$ 0.5 mmol/l | $\pm$ 0.09 |
| $K^+$ | 3.5 - 5 mmol/l | 9.2 | $\pm$ 0.1 mmol/l | $\pm$ 0.6 |
| $Ca^{2+}$ (free ionized) | 1.0 - 1.2 mmol/l | 2.3 | $\pm$ 0.02 mmol/l | $\pm$ 0.23 |

This accuracy of measurement not only makes great demands on electrodes and measuring equipment, it also sets certain requirements for the two calibration standards:
1) On the one hand, the concentrations in the two calibration standards should be close to the physiologically relevant range. On the other hand, large differences in concentration are of advantage because calibration lines derived therefrom should not to be too greatly influenced by the statistical measurement of error.
2) The ionic strength and electrolyte composition of the calibration standards should approach those of blood serum in order to keep the liquid junction potential differences at the reference electrode low.

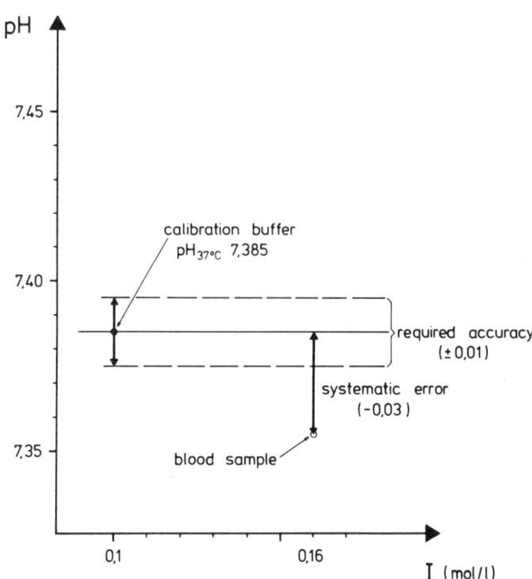


Fig. 1. Influence of ionic strength to pH-measurement in whole blood

3) The NBS (National Bureau of Standards) and DIN (Deutsche Industrie Norm)
have issued requirements for the preparation of solutions for the calibration of pH electrodes. In the pH range from 6.8 to 7.5 these buffers are
based on phosphate salts. These are the buffers normally used in clinical
chemistry. However, they have the disadvantage of having a lower ionic
strength of 0.10 mol/l in comparison to whole blood with 0.16 mol/l. This
results in a systematic error in blood pH determinations of ca. - 0.03 pH
units as compared to pH determinations with an electrode calibrated at 0.16
mol/l (cf. Fig. 1).
This systematic deviation has been described by Bates (1). Based on the
Henderson equation, the change in ionic strength from 0.10 to 0.16 mol/l
should produce a change of 0.01 pH units when a reference system of 3.5
mol/l KCL is used. On the other hand, the activity coefficient of the hydrogen ion alters, resulting in a pH change of ca. 0.006. The sum of these
two figures amounts to 0.016 pH units, whereas a mean deviation of 0.03 pH
units was in fact observed. This value clearly exceeds the required accuracy
of 0.01 pH units.

Table 2. Electrode signal changes introduced by changes in ionic strength

| ion M | $\Delta a_M$ (I: 0.145 $\longrightarrow$ 0.160 Mol/l) [ mV ] | normal range in blood serum or plasma [ mV ] |
|---|---|---|
| $H^+$ | 0.13 | 5.9 |
| $Na^+$ | 0.23 | 1.8 |
| $K^+$ | 0.25 | 9.2 |
| $Ca^{2+}$ | 0.38 | 2.3 |

Table 2 presents similar calculations for measurements with $Na^+$, $K^+$ and
$Ca^{2+}$ ion-selective electrodes. Shifts in ionic strength from 0.145 to 0.160
mol/l have been considered. By keeping the observed ion at a constant concentration, one can see significant changes in activity compared to the
range of normal values.

For the above-listed reasons, both standard solutions were optimised to an
ionic strength of 0.16 mol/l for the calibrations of $Na^+$, $K^+$ and $Ca^{++}$ and
pH electrodes. The activity coefficients for each ion in both solutions are
therefore approximately the same and can be ignored in blood determinations.

When preparing pH-buffered standard solutions for calcium it is also necessary to use a pH buffer system which does not form complexes with calcium.
It is therefore impossible to use phosphate - or carbonate-buffer systems.
The Tris buffer in common use today (= tris-(hydroxymethyl)-aminomethane)
has such a high pK value that its buffer capacity is severely limited in the
pH range of 7.0 - 7.4. The buffers HEPES (= N-2-hydroxyethylpiperazine-N-2-
ethanesulfonic acid) and TES (= N-tris(hydroxymethyl)methyl-2-aminoethane-

sulfonic acid) have considerably more favourable pK values. Since the sodium
salt of HEPES is strongly hygroscopic, whereas that of TES is not, the TES
system is to be preferred. Moreover, our studies have shown that TES does
not form complexes with calcium within the concentration range used.

Table 3. Two standard solutions for the simultaneous calibration of 5 different ion-selective electrodes for blood measurements

|  | standard solution 1 $[\text{mmol/l}]$ | standard solution 2 $[\text{mmol/l}]$ |
|---|---|---|
| NaCl | 115.0 | 75.0 |
| KCl | 3.0 | 7.0 |
| $CaCl_2$ | 0.8 | 1.5 |
| NaTES | 20.0 | 10.0 |
| TES | 14.16 | 24.89 |
| $MgCl_2$ | 1.53 | --- |
| $MgSO_4$ | --- | 12.125 |
| $NaN_3$ | 15.0 | 15.0 |
| pH ($37^\circ$C) | 7.385 | 6.839 |
| $Na^+$ | 150.0 | 100.0 |
| $K^+$ | 3.0 | 7.0 |
| $Ca^{2+}$ | 0.8 | 1.5 |
| $Cl^-$ | 122.66 | 85.0 |
| ionic strength | 160.0 | 160.0 |

Table 3 lists the composition of the two calibration standards together with
concentrations and pH values. The pH values were selected so as to correspond to those of the normally used phosphate buffers. There is however, a
systematic deviation of + 0.03 pH units in the pH determination of a blood
sample in comparison to conventionally calibrated devices. This discrepancy
must be tolerated if other ion selective electrodes have to be calibrated
with the same standard solutions. With the exception of the measurement of
pH, present-day medicine has not adopted activity as a measure for the other
ions. It is consequently necessary to calibrate ion-selective electrodes
with calibration standards of the physiological ionic strength. This ionic
strength is set using the magnesium salts referred to in Table 3. Sodium
azide is added in order to inhibit biological growth and to stabilize the
solutions.

These calibration standards make it possible to achieve adequate calibration
of ion-selective electrodes for blood analyses, although it does not totally
eliminate all measuring problems. Difficulties may, for instance, arise at
the liquid junction between blood sample and reference electrolyte. This

may be caused by suspension effects due to blood cells and proteins. It is also possible that proteins are salted out when highly concentrated electolyte solutions flow out of the reference electrode into the blood sample.

A further unsolved problem is the complexing of freely ionized calcium with anticoagulants, possibly combined with related dilution effects.

In conclusion, two calibration standards for the simultaneous calibration of five ion-selective electrodes for measurements in whole blood have been described. Two solutions having the physiological ionic strength of 0.16 mol/l are proposed which rule out the systematic error in pH measurement of present practice and permit the dertermination of the $\underline{Na}^+$, $\underline{K}^+$, $\underline{Ca}^{2+}$ and $\underline{Cl}^-$ ion concentrations in whole blood.

<u>References</u>

1. Bates RG, Vega CA, White DR Jr (1978) Standards for pH measurements in isotonic saline media of ionic strength I = 0.16. Anal Chem 50:1295

F. Hoffmann-La Roche & Co. Ltd. Bioelektronisches Dept., Grenzacher Str. 124, CH 4002 Basel / Switzerland

# Experience with Direct-Dip Potentiometry of Blood Serum Electrolytes

R.P. BUCK[*], H.D. SCHWARTZ[+]

A collaborative program with Applied Medical Technology, Inc. of Palo Alto, California has led to improved electrodes for direct potentiometry of blood serum electrolytes: $H^+$, $Na^+$, $K^+$, $Ca^{2+}$, $Cl^-$ and $HCO_3^-$. The basic instrument uses direct-dip of six electrodes (including a reference electrode) into discrete, 0.5 ml serum samples. Thirty samples and interposed pooled-serum standards move into the electrode station where they are sequentially measured.

The principal instrument, the AI-3 is an improved version of the original ionized calcium activity system (3). Features such as controlled atmosphere, controlled $CO_2$ partial pressure, self-balancing amplifier and direct read-out are retained. However, the system now uses software-control by micro-processor to determine sample position, electrode-dwell time, computation of slopes and intercepts from interposed pooled-serum standards, subsequent upgrading of calibration, computation of unknown concentrations and print-out.

Although electrodes measure ion activities, flame standardized pooled-sera are used as concentration standards. They are chosen to bracket the normal serum ion concentration range, and to lie close enough together (well within a factor of two) so that activity coefficient changes can be ignored. Consequently, calibration (mV vs concentration) is adequate, although sub-Nernstian. Referenced to aqueous standards, potentiometry of sera by ion-selective electrodes gives higher ion activities and concentrations (by about 7% on average) because of the excluded volume of protein materials. However, referenced to analyzed pooled-sera, flame and potentiometric data agree on average. The reference electrode uses a nearly equi-transferent electrolyte system to eliminate variability of liquid junction potentials. Bicarbonate is computed from measured pH, established $CO_2$ concentration and an average value for $pK_a$ of $H_2CO_3$.

Comparison of flame and potentiometric data have been published for $Ca^{2+}$, $Na^+$ and $K^+$ (2) and, more recently, for $Ca^{2+}$, $Na^+$, $K^+$, $HCO_3^-$ and $Cl^-$ (1). Correlation coefficients are better than 0.99 and slopes are within a few percent of unity. Statistical data have been published (1). Data on $HCO_3^-$ concentrations are remarkably good even though the quantity is calculated from measured pH and a value for dissolved $CO_2$ presumed to be valid for all sera. The system has also been evaluated for urine measurements and found to be suitable. Another instrument, the CI-3 is aimed at a two-channel (standard and sample) STAT-mode analysis.

Most recently, a line of solid-contact, coated-wire electrodes have been built. These electrodes supercede the membrane-configuration electrodes by avoiding the inner filling solutions. The new electrodes use essentially the same chemistries but provide a solid internal contact for the membranes. In addition, a new simple instrument model has been introduced for bedside, operation room and doctor's office settings. The newer electrodes are used in this simple, portable, direct-dip line of instruments for $Na^+$, $K^+$, $Ca^{2+}$ (or $Cl^-$). Controlled atmosphere is unnecessary and therefore eliminated. Response times are improved by rapid mixing using a vibrating sample holder.


## References


1. "AMT Clin-ion AI-3 system", J Clin Lab Instrum Reag (Japan) (1980) Issue 2, 3:85-93
2. Koellmann UT, Nelson JW (1978) Ionized calcium, sodium and potassium simultaneously by electrode. Clin Chem 366
3. Schwartz HD (1976) New techniques for ion-selective measurements of ionized calcium in serum after pH adjustment of aerobically handled sera. Clin Chem 22:461-467




*Department of Chemistry, University of North Carolina, Chapel Hill, N.C. 27514 USA

+Applied Medical Technology, Inc. 1059 E. Meadow Circle, Palo Alto, C.A. 94303 USA



## Discussion

Ahmad:  Could somebody explain the reactions taking place at the interphase between the membrane and the silver plate for a solid-contacted ion-sensitive PVC electrode?

Janata:  If the membrane/solid contact interface is non-polarized (i.e. resistive) its potential is uniquely characterized by the activity of the charged species which can cross it. Such an interface will tolerate a small net current without significantly changing its potential. On the other hand, if the membrane/solid contact interface is polarized (i.e. blocked, or capacitive) then the equilibrium is described by the activities of all species which interact with it; potential and charge. If the measuring instrument draws charge from this interface the equilibrium is disturbed and the interfacial potential will change. In order to avoid this situation, one should eliminate electrical leakage and use an electrometer with a high input impedance. In short, there is nothing wrong in principle with a membrane/solid contact interface in ISE's. One must be, however, careful about the selection of the electrical parameters of the entire measuring circuit. It is generally safer to use interfaces which approach the non-polarized behaviour.

# Intravascular K$^+$-Sensitive Electrodes for Clinical Monitoring

J.L. HILL

During the last few years we have witnessed almost an exponential rise in
the number of clinical monitors utilizing potassium-sensitive electrodes
based on the valinomycin-PVC matrix membrane (1, 3, 4, 5). The introduction
of these devices to the clinical setting has been limited, however, to in
vitro or extracorporeal measurements due to problems (or, at least, uncer-
tainties) concerning sterility and patient safety. We hoped to avoid some
of these problems by using a different polymer matrix that is biocompatible
and that can be sterilized; alleviation of these problems should permit direct on-line measurements of intravascular K$^+$ in man.


## Methods and Results

Potassium-sensitive electrodes were constructed from insulated silver wire
and valinomycin-polymer matrix membranes. The K$^+$-sensitive membrane was prepared according to the method of Le Blanc and Grubb (2). The membrane composition by weight was: 2.3% valinomycin (Sigma Chemical Co., St. Louis,
Missouri), 0.6% potassium tetraphenylborate (prepared from sodium tetraphenylborate, Sigma) and 97.1% poly (bisphenol-A carbonate), poly (dimethyl-
siloxane) block copolymer (Biochem International Inc., Wauwatosa, Wisconsin).
Silver wire (0.5 mm o.d.) was insulated with silicone polycarbonate (Biochem International Inc.). The insulation was stripped from 0.7 cm of each
end of a 13 cm wire; one end of the wire was chloridized, coated with approximately 0.1 mm of gelled isotonic KCl and dipped into a methylene chloride solution of the K$^+$-sensitive polymer. A luer-lock cap and electrical
connector were attached to the opposite end of the wire to complete the
sensor (Fig. 1). The electrode resistance ranged from 10 to 100 M$\Omega$. The EMF
response for a period of 10 months was linear from 2 to 20 mM K$^+$ and averaged
57 $\pm$ 1 mV per decade change in K$^+$ at 23$^\circ$ C. The 99% response time was less
than two seconds. Ion-selectivity was typical for valinomycin membrane electrodes (3). In addition, these characteristics were not altered by sterili-
zation with gamma radiation (2.5 - 3.5 MRads). Tests on materials for acute
toxicity and tissue reaction were negative (Science Associates, Northwood,
Ohio). The in vivo accuracy and stability was determined in 4 dogs and 8 pigs
anesthetized with $\alpha$- chloralose (70 mg/kg) and mechanically ventilated using
a Harvard pump. The K$^+$ electrode was inserted percutaneously into the femoral
vein (dogs) or jugular vein (pigs) through an 18 gauge teflon catheter (Ab-
bott Intracath Laboratories, N. Chicago, Ill.) with a plastic T-connector
at its exterior end. The K$^+$-sensitive tip extended into the vein 1.5 cm beyond the end of the catheter. An infusor unit (Sorenson Research Co., Salt Lake
City, Utah) was connected to the other part of the T-connector for slow infusion of sterile Ringers solution (3 ml/hr) to prevent clotting. In 4 experiments an additional infusor unit and a plastic telescope were attached
to the T-connector. This system permitted the withdrawal of the K$^+$ electrode
tip into the catheter lumen for in vivo calibration with two sterile Ringer
solutions of different K$^+$ concentration. A second catheter was inserted into the same vein for sampling of blood. A silver-silver chloride electrode

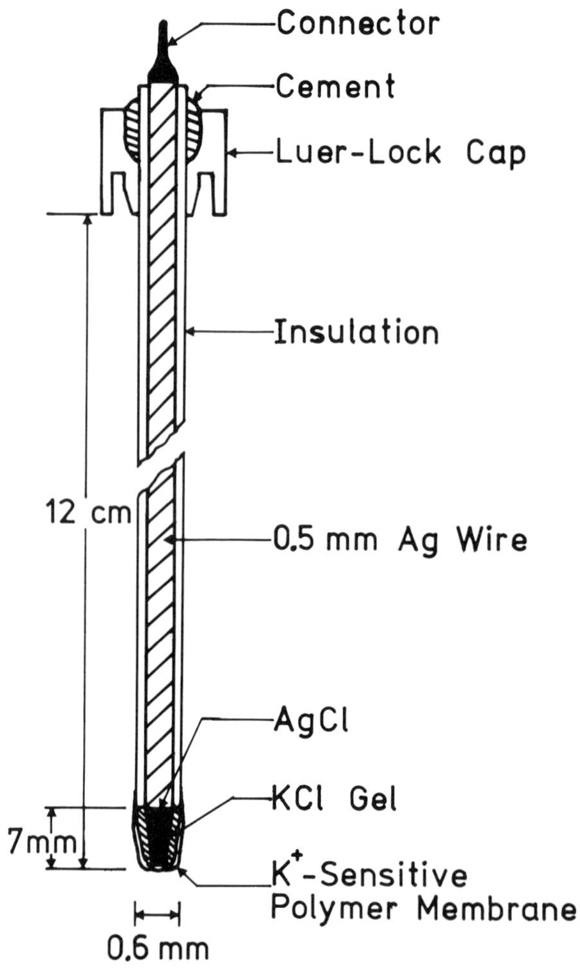



Fig. 1.  Construction of the intravascular $K^+$-sensitive electrode

containing a gelled chloride bridge (GE Daisy Patch) was positioned on the abdomen and used as the external reference electrode. EMF measurements were made using a battery-powered electrometer with electronics for direct $K^+$ readout, or a differential amplifier (Burr Brown 3670J); DC voltages were monitored on a digital voltmeter and continuously recorded on a stripchart recorder. Steady potentials were recorded within 5 min following electrode insertion.

Two methods for in vivo electrode calibration were employed. In the first method, a sample of venous blood was analyzed for $K^+$ by flame photometry; the steady-state electrode voltage was noted and equated to the measured $K^+$ concentration value. Thereafter, changes in electrode potential were converted to $K^+$ values by multiplying the original $K^+$ value by the antilog of the quotient of millivolt change and anticipated in vivo electrode slope.

The latter was computed from the in vitro slope and corresponding animal temperature using the Nernst relationship (4). These calculations were made electronically by the electrometer equipped with digital $K^+$ readout. In the second method, electrodes were calibrated by withdrawing the electrode tip into the lumen of the catheter during slow infusion of $K^+$ standards. Changes in $K^+$ were computed directly from this calibration line. This latter method provided a more analytical approach to in vivo determination of intravascular $K^+$, without dependence on other methods. Duplicate venous blood samples

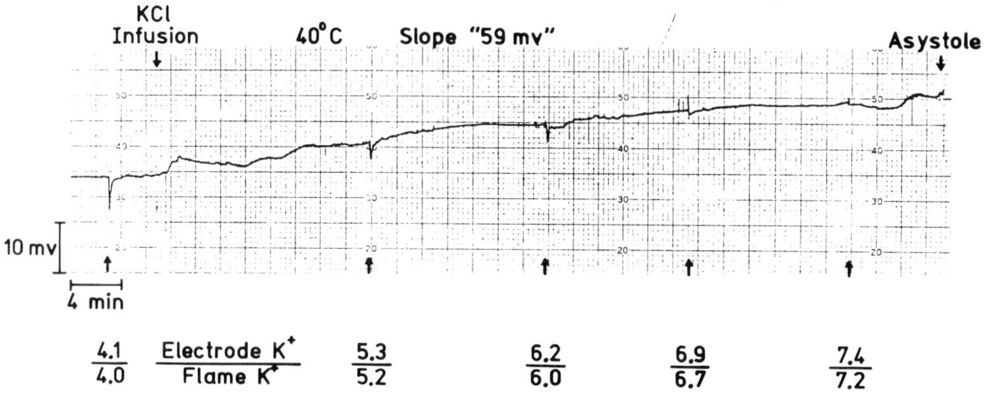


Fig. 2. Voltage tracing of changes in electrode EMF during venous $K^+$ elevation by continuous infusion of isotonic KCl. Arrows at bottom of tracing indicate points where blood samples were obtained. Numbers below arrows represent $K^+$ concentration (mM) determined by electrode and corresponding plasma values by flame photometry. The electrode was calibrated 2 hrs earlier using method one (described in text)

were drawn at 10 - 60 min intervals, heparinized, centrifuged and analyzed for plasma ($K^+$) by flame photometry (Instrumentation Laboratories Model 343). Venous $K^+$ was monitored by electrode and sampling methods for 2 - 6 hrs. In 8 animals the systemic $K^+$ level was elevated by slowly infusing KCl into the femoral vein. Fig. 2 illustrates an example of the changes in $K^+$ associated with KCl infusion. The $K^+$ electrode was calibrated by the first method described above, two hours before the beginning of this tracing. There was good agreement between electrode and flame values as the $K^+$ rose to 7.4 mM during the 1 hr period shown in this figure.

The results from 12 experiments comparing electrode and flame measurements are shown in Fig. 3. The venous $K^+$ concentration determined in vivo by electrode measurements correlated well to plasma $K^+$ values analyzed by flame photometry. The difference between electrode and corresponding flame measurements ranged from -0.1 to +0.3 mM in 16 determinations with a mean difference of 0.2 mM $\pm$ 0.1 (S.D.) for periods as long as 6 hrs without electrode recalibration (using either method). No evidence of thrombus formation on the electrodes was observed.

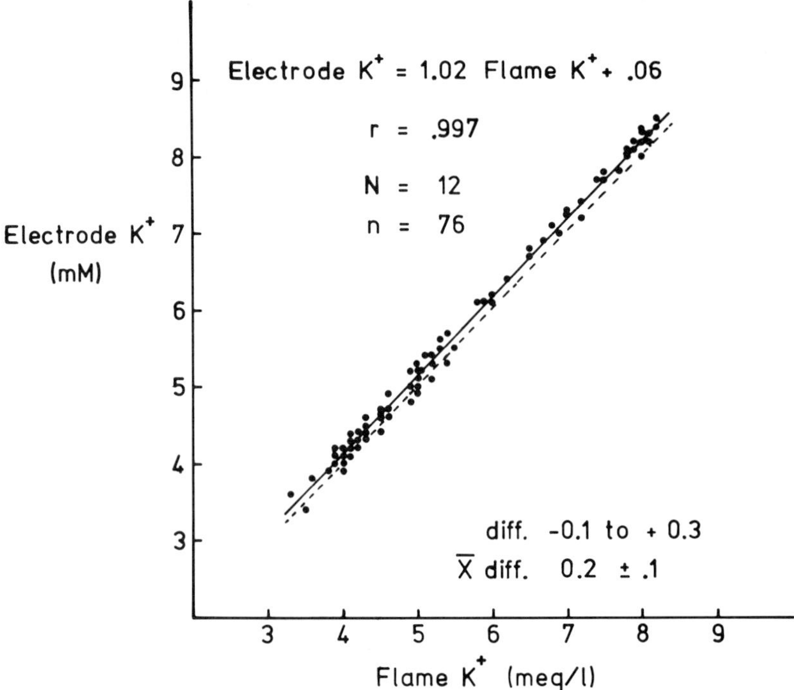



Fig. 3. Comparison of electrode and flame K$^+$ measurements by regression analysis. The dashed line represents the line of identity.


Summary

The results from these studies indicate that coated wire electrodes made with valinomycin and these special silicone copolymers may provide accurate and safe in vivo measurement of intravascular K$^+$ in the clinical setting. These electrodes are easily inserted and calibrated in vivo; and with appropriate electronics these methods provide direct readout of K$^+$ concentration values for instantaneous and trend monitoring. These electrodes can be made in different diameters, lengths and configurations for various clinical applications.


References


1. Ammann D, Jenny HB, Anker P, Oesch U, Simon W (1980) Carrier based ion-selective liquid membrane electrodes and their medical applications. (This symposium)
2. Le Blanc OH, Grubb WT (1976) Long-lived potassium ion-selective polymer membrane electrode. Anal Chem 48:1658-1660
3. Meier PC, Ammann D, Osswald HF, Simon W (1977) Ion-selective electrodes in clinical chemistry. Med Prog Technol 5:1-12

4. Osswald HF, Asper R, Dimai W, Simon W (1979) On-line continuous measure-
     ments of potassium concentration in whole blood during open-heart sur-
     gery. Clin Chem 25:39-43
5. Treasure T (1978) The application of potassium selective electrodes in
     the intensive care unit. Intens Care Med 4:83-89


## Acknowledgements

This work was supported in part by U.S.NHLBI grant HL 23624. I would like
to express my appreciation to Biochem International Inc., Wauwatosa, Wisconsin for supplying the silicone copolymer materials and battery powered
electrometer.


Department of Medicine, Division of Cardiology, University of North Carolina
School of Medicine, Chapel Hill, North Carolina/USA

# Clinical Evaluation of a Nova 1 Sodium/Potassium Electrode in a Cardiac Surgical Intensive Care Unit

T. TREASURE, W. AVELING, I.W. BROWN, E. CORNWELL, D. O'CONNOR, J. SIMPSON
K.J. WARK

## Introduction

During cardiopulmonary by-pass and in the immediate postoperative period, serum potassium can change quite rapidly, usually tending to fall. Unless this deficit is corrected by potassium replacement at an appropriate rate there may be difficulty in defibrillating the heart in order to discontinue by-pass or there may be dangerous postoperative arrythmias. Frequent measurements of potassium should therefore be available for the safe management of these patients and improved techniques for their provisions are of great interest to the surgical teams responsible for the treatment of these patients.

When the emission flame photometer is used, sodium and potassium are usually measured on the same sample and the two values are reported together. In the context just described, where the potassium may be requested several times within an hour, the sodium measurement is of little relevance. Information is required immediately and frequently about potassium and similarly, for arterial $pO_2$, $pCO_2$ and pH since these can change rapidly and must be corrected promptly if disaster is to be avoided. On the other hand, sodium measurements, along with urea and creatinine for example, are not usually needed more often than once a day. With the advent of new technology there seemed to be an  opportunity to reconsider clinical requirements and make analyses in more logical groupings rather than always measuring sodium and potassium together for traditions sake. However the commercially available systems have continued to package the two together but for the clinical reasons stated we have largely confined our study to the potassium measurements of NOVA 1.

In 1970, Simon's group reported the use of valinomycin in a liquid membrane to measure serum potassium (4) and the subsequent development of electrodes with PVC membranes permitted estimations to be made on heparinised but otherwise untreated specimens of whole blood (3, 6). It was quickly realised that a potassium ion selective electrode sited in the Intensive Care Unit would be a convenient way of providing the rapid potassium estimations that are needed. Confirmation of a simple and predictable relationship between activity and concentration measurements of potassium in serum and whole blood (1) permitted us to introduce potassium electrode measurements into routine postoperative care (5). We have also considered the possibilities of "on-line" intravascular measurements (7) and we reported our early experience with continuous in-vivo potassium measurements at the IUPS Satellite Symposium in Dortmund in 1977 (8). However, we remain of the opinion that while catheter tip electrodes are valuable research tools (2), a reliable bench electrode, sited near the patient is the safest way to monitor for potassium therapy in routine postoperative care (5). This paper therefore deals only with in-vitro electrode measurements on whole blood samples.

Our previous experience with electrode measurements of potassium for clinical purposes had been with valinomycin/PVC membranes of our own construction which were manually operated and calibrated (1, 5, 6). That 'home made' electrode system has remained in continuous service in the unit where it was first studied. The very great advantages of having a potassium measurement available within two minutes of taking a sample of blood have resulted in the electrode becoming virtually indispensible. In March 1980 a commercially produced, automated, self-calibrating sodium/potassium analyser (NOVA 1) was installed in the cardiac surgical unit of The London Chest Hospital thus making available rapid potassium estimations on all patients in the operating theatre and intensive care unit. Prior to that date all potassium estimations had been provided by the central laboratory using an emission flame photometer.

We have evaluated the NOVA 1 during the first three month period after its installation and compared it with our existing laboratory service. We already know form our own experience and that of others that there is good correlation between electrode and flame photometer measurements in carefully performed laboratory comparisons. The purpose of this study was to see if a self-calibrating electrode system in the hands of non-technical staff would provide us with a service as reliable and accurate as the central laboratory and, if in fact, it proved to be quicker and more convenient.


## Materials and Methods

The NOVA 1 sodium/potassium analyser was brought into service during March 1980 and was studied for a period of just over three months, from the time of its installation until the planned conclusion of the trial period at the end of June 1980. The machine was used exactly according to the suppliers instructions but the trial was not arranged in conjunction with the manufacturers or agents. The servicing facilities were those routinely available to other users of this equipment. We believe this is important in a realistic evaluation of a machine of this type which has been specifically designed for use by non-technical personnel. As has already been explained, we are much more interested in the potassium measurement than sodium because of the relative urgence of the information so we confined our interest to the potassium channel almost exclusively.


## NOVA 1: Function and Method of Use

The NOVA sodium and potassium analyser uses ion selective electrodes which we believe to be valinomycin in a polymer membrane for potassium and a glass electrode for sodium. It automatically performs a two point calibration every two hours and during standby, minute amounts of calibration solutions are slowly pumped through the analyser so that the tubing is kept free flowing and the electrodes conditioned. The machine is thus ready for immediate use with no warm up period or manual calibration required. Heparinised whole blood is aspirated by the machine so that it controls its own sample volume (0.3 ml) and flow rate. A single point calibration is performed with each analysis to check for drift. The result is displayed within a minute in a digital fashion. The whole process is controlled by a microcomputer which can also detect, and to some extent analyse faults. Measurements which are detected as likely to be erroneous are not displayed. Instead a "status code" is displayed which suggests the nature of the problem and speeds diagnosis and therefore helps in "trouble shooting".

Our study was aimed at establishing:

1) The service reliability of the machine.

2) The day to day precision and reproducibility of the electrode measurements as determined with external standards.

3) The correlation betwen NOVA 1 and the service pathology laboratory potassium measurements.

## Service Reliability.

A daily log was kept, recording the performance of the machine and any faults or problems which developed. This included the exact nature of the fault and the time which elapsed before the analyser was functioning satisfactorily again (so called "down time").

## Precision on External Standards.

Ion activity standards containing 2.0 and 4.5 mM potassium chloride in a background of 140 mM sodium chloride (1) were used twice daily as an external check on the electrodes precision, stability and reproducibility. Up to 12 individual measurements on standards were made on some days to test within day consistency and a total of 515 external calibrations were made over a 104 day period.

## Correlation with Flame Photometer.

During the trial period part of the sample of heparinised blood drawn from the patients in the operating theatres or in the postoperative intensive care unit was put through the NOVA 1 and the remainder sent to the pathology laboratory for routine measurements. Both measurements were performed singly by which ever member of the hospital staff would normally be responsible. The results were recorded independently. The consultant pathologist was aware that the electrode was undergoing evaluation in the unit but by agreement no attempt was made to indentify the paired specimens from the routine throughput of the laboratory. Our intention was to test the NOVA 1 under exact working conditions with measurements made singly by any one of a number of responsible but not specially trained staff and to compare the results with our routine laboratory service.

## Results

### 1) Service Record

The period during which the machine was evaluated totalled 104 days. During that time the NOVA 1 was out of action on three occasions. All three faults were minor and related to the mechanics of fluid transport through the machine. However, they resulted in a total of 11 days "down time". The individual times out of service were 1, 4 and 6 days.


## 2) External Standards

The within day precision of the electrode was very high. For example a series of repeated calibrations with 4.5 mM potassium chloride activity standard produced the following result:

    n = 11
    mean of potassium measurements = 4.48 mM
    s.d. = 0.186
    95% extreme range = 4.44 - 4.52 mM

The full series of 515 external calibrations over the 104 day test period were also analysed. This included 2.0 and 4.5 mM standards and the results are given as a discrepancy from the normal value:

    n = 515
    mean of sodium measurements = 140.02 mM
    s.d. = 0.28
    95% extreme range = 139.46 - 140.58 mM


## 3) Correlation with Flame Photometer

A total of 692 blood samples were analysed both by the ion selective electrode and the pathology laboratory flame photometer. The majority of the estimations fall between 4 and 5 mM potassium because of our policy of controlling potassium very carefully. Therefore the most useful way of displaying the information graphically is to give the discrepancy between the two readings. This is calculated as the electrode result minus the flame photometer result (ISE - EFP) so that the discrepancy is the deviation of the electrode measurements from that of the flame photometer. These results are plotted as a discrepancy frequency histogram (Figure 1) so that all 692 measurements are recorded. They are separated into three groups according to calendar month so that any learning effect can be seen. The results for the last one month period, which included 242 comparisons are displayed in Figure 2 where the mean discrepancy with its standard deviation is given. The regression analysis for these 242 comparisons gave a regression coefficient of $r = 0.815$ and the equation for the regression line was:

$$y = 0.15 + 1.02x$$


## Discussion of Results

During the period when the NOVA 1 was on trial we performed major cardiac surgery on 128 patients and used the electrode to provide us with 1700 rapid potassium estimations used in the treatment of these patients. All the staff involved were impressed with the speed and simplicity of operation of the analyser. Clinical staff appear to have complete trust in the measurements produced by either the laboratory or a 'black box' in their unit so this favourable impression must be converted into hard evidence before we rely completely on new machinery introduced into the care of our patients. Furthermore, it is not enough that the system works

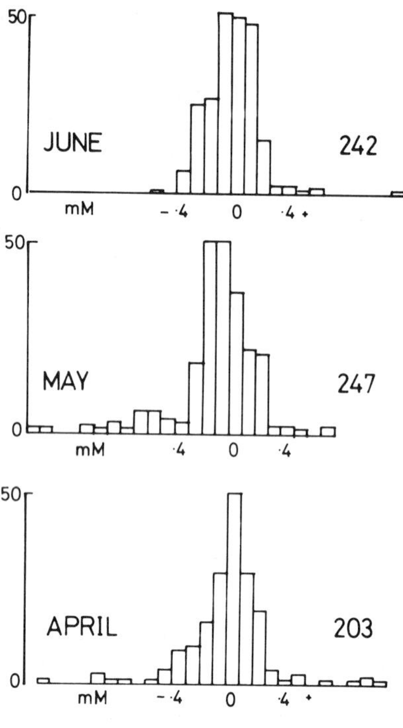



Fig. 1. Six hundred and ninety two comparisons of potassium measurements made with NOVA 1 in the intensive care unit and an emission flame photometer in the service laboratory. The discrepancy has been calculated as ISE-EFP and plotted as a frequency histogram for three separate calendar months

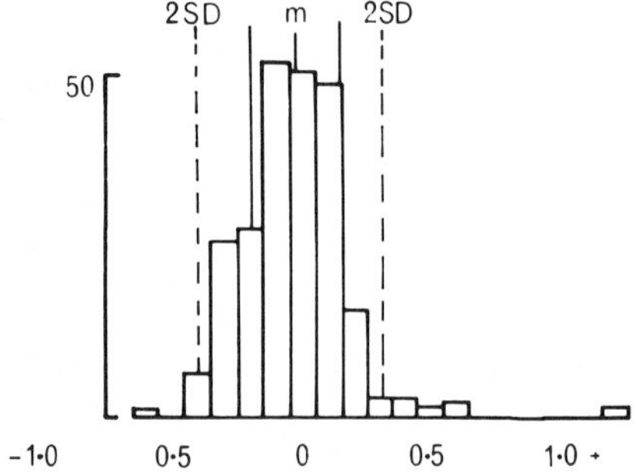



Fig. 2. Two hundred and forty two comparisons during the third month of evaluation. The results have been displayed as in Figure 1 and the deviation of the electrode measurement from the flame photometer is given as +/- mM in a frequency histogram. The mean discrepancy was −0.04. 95% of the extreme range of the discrepancy for this month was −0.32 – +0.4 mM

in the careful hands of its inventors; it must also stand up to less experienced and perhaps more critical scrutiny.

The analyser developed significant faults on only occasions during the 104 days trial period. That in itself seems a very satisfactory service record but the fact that the analyser was out of commission for a total of 11 days as a result of these few faults was less acceptable. None of the faults would have taken more than a few minutes to fix by an experienced technician, familiar with the NOVA 1 and armed with a few spare parts. The possible solutions to this problem are obvious and do not merit further discussions but it is clear that any unit contemplating the purchase of equipment of this type, which is to be used outside the pathology department, should make provisions for the routine maintenance of the equipment and allow for the necessary staff time for this to be done. This has become the policy in our unit, but it leaves the problem of out of hours and holiday cover.

The within run calibration was extremely good with most of the series of measurements falling within an extreme range of 0.05 mM. The day to day calibration was also excellent with 95% of 515 values falling within +/- 0.14 mM of the nominal standard even though these measurements were spread over a period of over three months. This very precision is more than adequate for clinical purposes where a result to 0.2 mM is accepted as satisfactory. These measurements on the aqueous external standards were made by one of two observers with experience of analytical equipment. While the results amply confirm the precision and stability of the electrode it is the analysis of blood samples taken to the machine by up to twelve different non-technical staff that confirmed its value in practice.

The results of the 692 correlations with the flame photometer demonstrate first of all that there is a learning curve with less outlandish points seen in the third month than in the previous two. It must also be said that a study performed in this way gives the worst possible case. A variety of clinical and technical staff were involved on both sides. No corrections can be made for clerical, transcription or telephone errors in the communication and recording of the results which must have occurred somewhere in the course of 692 comparisons. All the measurements were made singly because our routine is to only repeat if the result is widely abnormal or in some other way unexpected. Finally, the method of displaying the results in Figures 1 and 2 does not permit any occasional extremely discrepent result to become lost amongst a statistical analysis where it has little influence on an otherwise satisfactory mean and standard deviation. On the discrepancy frequency histogram every single data point is displayed. This is important because a single false value may have disastrous consequences. Patients can die both because of a failure to diagnose a low potassium and correct it or because a rising potassium is not identified in time and an extra lethal infusion of potassium chloride may be given. We have managed to control potassium within safe margins using our existing services. Any replacement must be at least as safe as that. Our evidence is that NOVA 1 produces accurate results and can be relied on to not display a result which could be misleading due to internal errors.

## Conclusions

NOVA 1 has lived up to all the claims made for it by its manufacturers. It provides accurate and reliable results for potassium on whole blood within a minute. It is available for use virtually at the bedside and the speed and quality of the service is of course equally good throughout twenty-four hours. The only problems we have encountered are related to maintenance and servicing and it is clear that if analytical equipment is to be used outside the immediate supervision of the pathology laboratory proper provision must be made for its upkeep.

## References


1. Band DM, Kratochvil J, Poole-Wilson PA, Treasure T (1978) Relationship between activity and concentration measurement of plasma potassium. Analyst 103: 246
2. Band DM, Linton RAF, Treasure T (1979) Demonstration of potassium electrode suitable for use in vivo in man. J Molec & Cell Cardiol 11 (suppl 2): 2
3. Moody GJ, Oke RB, Thomas JDR (1970) A calcium sensitive elcetrode based on a liquid ion exchanger in a PVC matrix. Analyst 95: 910
4. Pioda LAR, Simon W, Bosshard HR, Curtius HC (1970) Determination of potassium ion concentration in serum using a highly selective liquid-membrane electrode. Clin Chim Acta 29: 289
5. Treasure T (1978) The application of potassium selective electrodes in the intensive care unit. Intensive Care Medicine 4: 83
6. Treasure T, Band DM (1977) Measurement of plasma potassium using ion selective electrodes. Proc Anal Div Chem Soc 14: 334
7. Treasure T, Band DM (1977) A catheter tip potassium selective electrode. J Med Eng & Techn 1: 271
8. Treasure T, Band DM (1978) Continuous measurement of plasma potassium with a catheter tip ion-selective membrane. Arzneim Forsch (Drug Res) 28: 881



Departments of Surgery and Anaesthesia, The London Chest Hospital, Bonner Road, London E2 9JX

# Distribution of Intra- and Extracellular K$^+$ in the Leech Central Nervous System Studied Using Double-Barrelled Ion-Sensitive Microelectrodes

J.W. DEITMER[*], W.R. SCHLUE[+]

## Introduction

The maintenance of an ionic homeostasis within the central nervous system
is undoubtedly essential for proper neuronal functioning in both vertebrates
and invertebrates (9, 10). The leech central nervous system has been the
subject of many electrophysiological studies at cellular and behavioural
levels (2). In the present study we have used double-barrelled K$^+$-sensitive
liquid ion-exchanger microelectrodes to measure intracellular (in the Retzius cell somata) and extracellular K$^+$ activity; also to monitor changes in
K$^+$ produced by altering the [K$^+$] of the bathing solution and by the addition
of the cardiac glycoside ouabain (C-strophanthin).

## Methods

Single ganglia of the leech _Hirudo medicinalis_ were used for experimentation.
Details of the dissection and the preparation have been described previously
(8). The normal bathing solution contained (in mM): NaCl, 115; KCl, 4; CaCl$_2$,
1.8; Tris-Maleate, 10 (pH 7.4); glucose, 11. When the [K$^+$] of the bathing solution was reduced or increased, equivalent amounts of Na$^+$ were added or removed. Ouabain (Serva) was added solid to the solutions to give a final concentration of 5x10$^{-4}$ M. This amount of ouabain had no effect on the response
of the electrodes. The preparations were continuously superfused with a flow
rate of 15 to 20 bath volumes/min (bath volume _ca._ 0.2 ml). The experiments
were performed at room temperature (22 to 25$^{\circ}$ C). The construction of the
double-barrelled K$^+$-sensitive microelectrodes have been described in detail
elsewhere (8). The tip of one barrel was filled with a small column of K$^+$-
sensitive liquid ion-exchanger (Corning 477317), the rest of the shaft was
filled with 0.5 M KCl. The other barrel was filled with either 3 M sodium
acetate or 1 M magnesium acetate, and served as a reference electrode.
The K$^+$-sensitive microelectrodes used responded with a potential of 45 to
52 mV (mean 49 $\pm$ 1.8 mV, n = 34), when the [K$^+$] was changed from 4 to 40 mM.
Above a [K$^+$] of 10 mM the mean electrode response increased by up to 3 mV.
The selectivity coefficient for Na$^+$ over K$^+$, $k$, as determined by the Nicolsky
equation was 1/62. The response time of the electrodes ($\lesssim$ 1 $\mu$m tip) was usually 200 to 500 ms for a half maximum response. The DC resistance of the
electrodes was 1 to 5x10$^9$ $\Omega$ for the K$^+$-sensitive barrel, and 1 to 5x10$^7$ $\Omega$ for
the reference barrel. The preparation and the recording system are shown in
Fig. 1.

## Results and Discussion

Fig. 2 shows a profile of the potentials measured by the reference and the
K$^+$-sensitive microelectrodes when the electrode is advanced into an intact
ganglion. First the electrode was pushed through the outer ganglion capsule

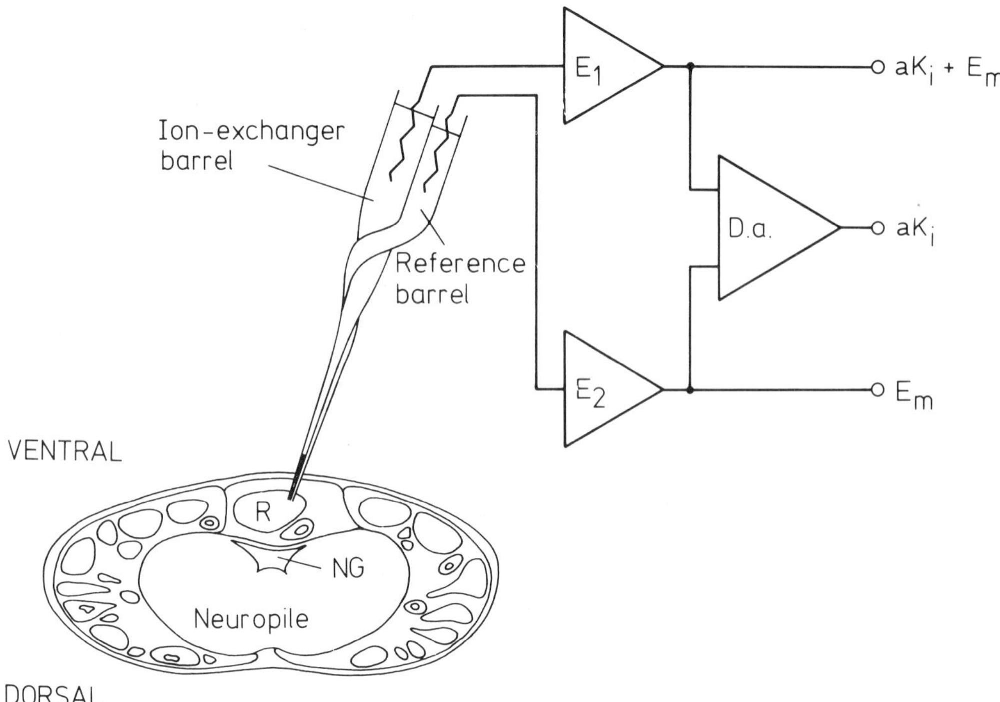


Fig. 1. Schematic drawing of the preparation and the electrical recording system. The lower part of the figure shows a ganglion cross section with the ventral side upwards (R, Retzius cell; NG, neuropile glial cell). The ion-exchanger barrel and the reference barrel of the double-barrelled micro-electrode are each connected via chlorided silver wires to two independent electrometer probes ($E_1$ and $E_2$, input resistance $10^{15} \Omega$, bias current less than $10^{-14}$ A). The probe outputs were fed into the inputs of a differential amplifier (D.a., WPI F-223A). The signal from the differential amplifier, giving the pure $K^+$ activity measurement, $\underline{a}K_i^+$, and the signal from probe $E_2$, giving the cell membrane potential, $E_m$, were connected to two independent channels of a pen-recorder

and the endothelial layer into a Retzius cell body. Then the electrode was slowly withdrawn until a sudden jump in the traces indicated an extracellular recording. The reference potential was near zero, but the $K^+$ level in this extracellular space was still significantly higher than in the bath. After the same Retzius cell was impaled again, the electrode was pushed through this cell until its tip emerged at the other side into the extracellular space. This extracellular space also contained a higher $K^+$ level than in the bath. These two recording sites, situated in the extracellular spaces of the nerve cell body region indicated a mean $[K^+]$ of $5.8 \pm 0.6$ mM (S.D., n = 27).

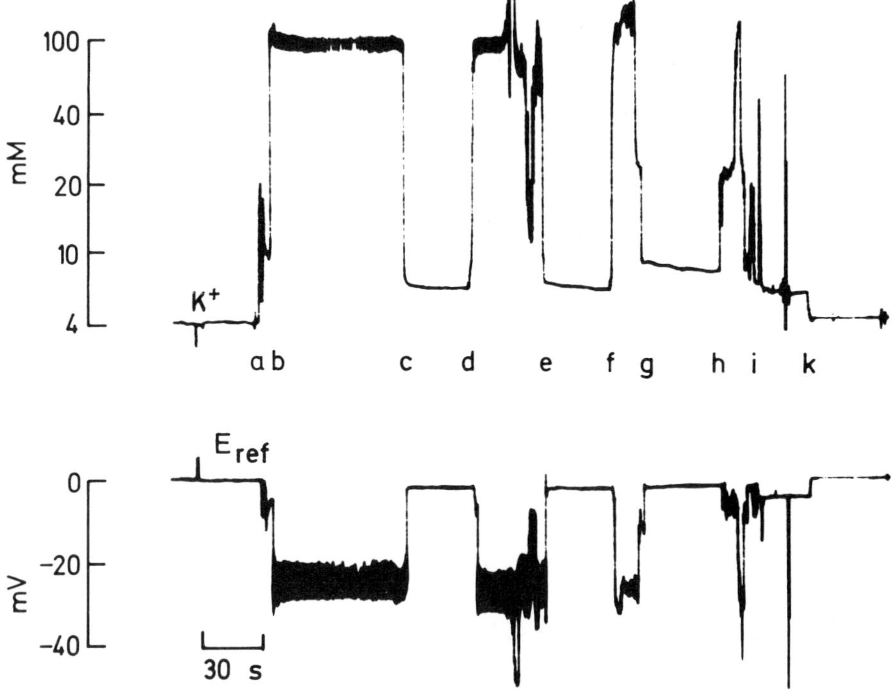

Fig. 2. Profile of the potassium concentration ($K^+$) and the potential
($E_{ref}$) recorded as a double-barrelled microelectrode penetrated a single
leech ganglion from the ventral surface. The lower case letters indicate
the position of the electrode tip within the ganglion, as follows: (a) passage through the endothelium and the outer ganglion capsule; (c) entry into a first extracellular space in the nerve cell body region, defined by
the penetration of, and subsequent withdrawal from, a Retzius cell during
the time between (b) and (c) (the recording indicates that the potassium
concentration here is higher than in the bathing medium); (d) repenetration
of the same Retzius cell; (e) subsequent passage into a second defined extracellular space of the nerve cell body region; (f) passage through the
inner ganglion capsule and entry into a cell not absolutely indentified,
(possibly the anterior neuropile glial cell, which, being injured by the
electrode, has a potential difference less than would be expected for an
intact cell, only ca. -30 mV); (g) emergence from the unidentified cell
and entry into the neuropile (as in the nerve cell body region, the potassium concentration here is higher than in the bath); (h) deeper penetration
into the neuropile (indicated by repeated injury discharges of pierced
axons); (i) withdrawal into the nerve cell body region (presumably the reference barrel was now blocked); (k) withdrawal from the ganglion into the
bathing medium

When the electrode was advanced even further into the ganglion, it passed
the inner ganglion capsule and briefly penetrated an unidentified cell, presumably the anterior neuropile glial cell (see Fig. 1 for orientation). However, this cell was probably damaged and could not be maintained for a stable
recording. The electrode emerged on the far side of this cell into another
extracellular space, situated in the neuropile. The $K^+$ level in the extracellular spaces of the neuropile was even higher, on average $6.3 \pm 0.7$ mM
(S.D., n = 15). Upon advancing the electrode deeper into the ganglion, injury discharges of pierced axons were often recorded, confirming that the
recording site was within the neuropile.

The mean $K^+$ activity measured in the Retzius cell bodies was $101.3 \pm 7.6$ mM
(S.D., n = 14). Assuming an intracellular activity coefficient for $\overline{K^+}$ of
about 0.75 (6), this would indicate an intracellular $[K^+]$ of 135 mM. The
mean membrane potential of these Retzius cells was $-43.6 \pm 4.9$ mV. With the
extra- and intracellular $K^+$ measured, the $K^+$ equilibrium potential of the
Retzius cells was calculated according to the Nernst equation to be $-80$ mV.
This is some 36 mV more negative than the mean membrane potential, indicating that $K^+$ ions are actively accumulated in these neurones against a considerable electrochemical gradient.

The change of the $K^+$ in the extracellular spaces of intact ganglia, following increase or decrease in the $K^+$ of the bathing solution, occurred exponentially. The half times were 20 to 30 s for both the $K^+$ changes in the
neuropile and in the nerve cell body region. When the outer ganglion capsule
was opened, these half times were much reduced, i.e. to between 3 and 10 s.
It appears that the outer ganglion capsule is a diffusion barrier for $K^+$ to
move into and out of the intact ganglion.

The steady-state $K^+$ levels in the extracellular spaces of intact leech ganglia at various $[K^+]$ of the bathing solution is shown in Fig. 3. At all $[K^+]$
in the bath tested, the $K^+$ level in the extracellular spaces of the neuropile and nerve cell body region was higher than in the bath, in particular
when the $[K^+]$ in the bath was low (<4 mM). In the absence of $K^+$ in the bath,
there was still a significant amount of $K^+$ maintained in the extracellular
spaces. For all bath $[K^+]$, except 0.2 and 2.0 mM, there was a significant
difference (p<0.05 by the t-test) between the steady-state values of the nerve
cell body region and of the neuropile. When the outer ganglion capsule was
carefully disrupted and opened, the $K^+$ level measured in the ganglia was the
same as in the bath. The tip of the double-barrelled microelectrode was advanced deeply into the ganglia in these experiments, and its site of recording was probably within the extracellular spaces of the neuropile. The broken
line in Fig. 3 indicates identical $K^+$ levels in the opened ganglia, as compared with the bath. Whether this was due to disruption of the outer capsule
and endothelial layer, or to the widening of the extracellular cleft system
remains unresolved. It appears, however, that the $K^+$ level within the extracellular spaces of _intact_ ganglia is maintained high by passive and/or active
processes.

A possible source of the elevated $K^+$ level in the extracellular spaces of
intact ganglia are nerve and/or glial cells. We have measured the intracellular $K^+$ activity of Retzius cell bodies at a high and a low $[K^+]$ in the
bath. When the $[K^+]$ in the bath was raised from 4 to 40 mM, there was little
or no change in the intracellular $K^+$ activity. When the $[K^+]$ of the bathing
solution was reduced to zero, the intracellular $K^+$ activity began to de-

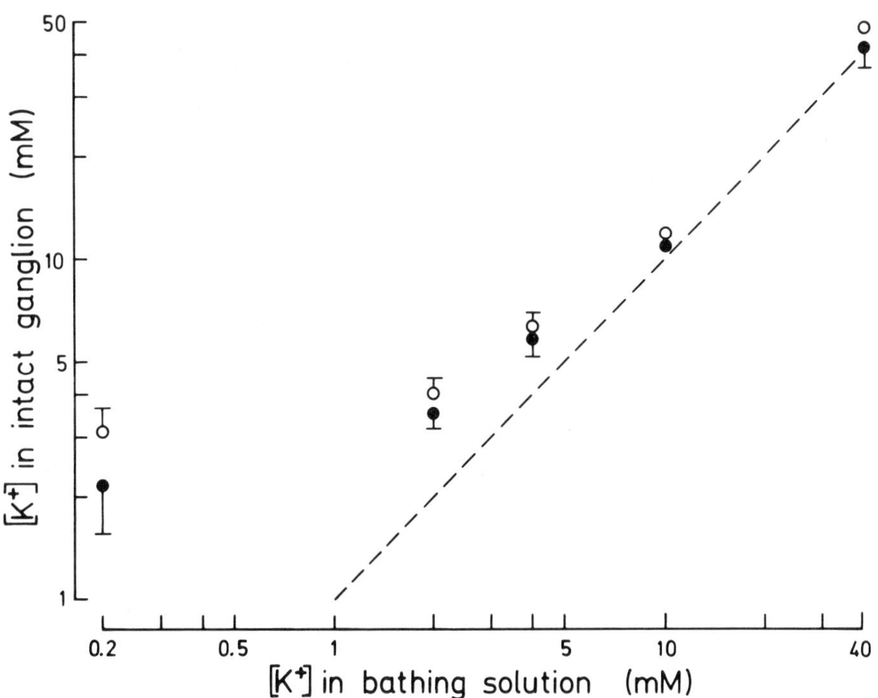

Fig. 3. The $K^+$ concentration as measured with the double-barrelled $K^+$-sensitive microelectrode in the extracellular spaces of the nerve cell body region (filled circles), and of the neuropile (open circles) at various $K^+$ concentrations in the bathing solution (between 0.2 mM and 40 mM). The dashed line indicates equal $K^+$ concentrations in the extracellular spaces and the bathing solution (which were measured when the outer ganglion capsule and the endothelial layer were opened). The vertical bars indicate + or − one standard deviation

crease within 3 to 5 min. The maximum rate of decrease of the intracellular $K^+$ activity during the next 10 min was up to 12 mM/min. The membrane <u>depolarized</u> upon removal of $K^+$, although an increase of $K^+$ in the extracellular spaces under these conditions never occurred. When the $[K^+]$ in the bath was restored to 4 mM, the intracellular $K^+$ activity rapidly increased again, the membrane hyperpolarized beyond the initial level, and then slowly depolarized again. This transient hyperpolarization increased in amplitude and duration with the duration of the exposure to zero $[K^+]$, and was independent of the time course of the rise of $K^+$ in the extracellular spaces. There was also no depletion of $K^+$ observed in the extracellular spaces of the nerve cell body region when the $[K^+]$ of the bath was restored. These results suggest that the membrane potential changes upon removal of $K^+$ might be due to inhibition and re-activation of an electrogenic Na-K pump in the Retzius cell membrane.

98

In order to test this hypothesis, we applied the cardiac glycoside ouabain
($5 \times 10^{-4}$ M) to the bathing solution. The $K^+$ levels in the extracellular spaces _transiently_ increased following the addition of ouabain. This increase
amounted to 1.2 $\pm$ 0.4 mM (S.D., n = 5) in the nerve cell body region, 3.8
$\pm$ 1.0 mM (S.D., n = 6) in the neuropile, and was little changed when the
ganglion capsule had been opened. In all experiments the $K^+$ levels returned
to their initial states, even in the continuous presence of the glycoside.

The intracellular $K^+$ activity of the Retzius cells began to decrease rapidly
within 2 to 3 min after the addition of ouabain. The maximum rate of decrease
was up to 14 mM/min. The cell membrane depolarized in several phases during
this time. The intracellular $K^+$ activity continued to fall to about 10 mM,
while the membrane potential decreased to -10 mV within 10 to 15 min of ouabain application. Taking the maximum rate of decrease of the intracellular
$K^+$ activity as the rate of _net_ $K^+$ efflux from the cells, and a surface/
volume ratio of $7.5 \times 10^2$ $cm^{-1}$ (3), this $K^+$ efflux would amount up to 0.23 mM/
$cm^2$s in the presence of ouabain, indicating an 'apparent' $K^+$ permeability
coefficient, $P_K$, of $8 \times 10^{-6}$ cm/s for the Retzius cell membrane.

The high $K^+$ levels in the extracellular spaces of intact leech ganglia might
be influenced by passive loss and/or active reuptake of $K^+$ ions from and
into nerve and glial cells. In glial cells of this preparation, the $K^+$ distribution across the cell membrane has been reported to be in equilibrium,
in contrast to the nerve cells (4, 5). Our results suggest that the intercellular cleft system and/or the outer ganglion capsule may act as diffusion
barriers. A part played by the endothelial layer, situated just next to the
outer capsule, e.g. by actively pumping $K^+$, cannot yet be excluded. Other
mechanisms, such as surface charge effects at cell membranes and the presence
of negatively charged polyelectrolytes in the intercellular fluid, may also
produce a high $K^+$ level in the extracellular spaces. We have already discussed the possibilities of damage to neurones and glial cells, and experimental artefacts, as the cause for the high $K^+$ levels in the extracellular
spaces (8); but these are unlikely to produce a _sustained_ high $K^+$ level.
In contrast, it seems that the $K^+$ level within these narrow spaces may in
fact be even higher than measured with the double-barrelled microelectrodes
since artificially large spaces are created around the electrode tip.

The regulation of the $K^+$ level in the extracellular spaces of the nerve cell
body region and the neuropile might be important in modulating some processes of integration within the central nervous system of the leech. The spontaneous and evoked discharge of neurones, and the occurrence of synaptic
potentials have been shown to be affected by altered $[K^+]$ in the bath (1).
Accommodation of some sensory nerve cells also depends on the extracellular
$K^+$ level (7). The maintenance of high intracellular and extracellular $K^+$
levels in the intact central nervous system of the leech might thus play an
essential role in providing the conditions for normal neuronal functioning.


<u>References</u>


1. Baylor DA, Nicholls JG (1969) Changes in extracellular potassium con-
     centration produced by neuronal activity in the CNS of the leech.
     J Physiol 203:555-569
2. Kuffler SE, Nicholls JG (1976) From neuron to brain. Sinauer Assoc
     Inc Publ, Sunderland Mass USA

3. Lent CM (1977) The Retzius cells within the central nervous system of leeches. Progr Neurobiol 8:81-117
4. Nicholls JG, Kuffler SW (1964) Extracellular space as pathway for exchange between blood and neurons in the CNS of the leech. Neurophysiol 27:645-671
5. Nicholls JG, Kuffler SW (1965) Na and K content of glial cells and neurons determined by flame photometry in the central nervous system of the leech. J Neurophysiol 28:519-525
6. Robinson RA, Stokes RH (1959) Electrolyte solutions. Butterworth, London
7. Schlue WR (1976) Sensory neurons in leech central nervous system: changes in potassium conductance and excitation threshold. J Neurophysiol 39:1184-1192
8. Schlue WR, Deitmer JW (1980) Extracellular potassium in neuropile and nerve cell body region of the leech central nervous system. J Exp Biol (in press)
9. Somjen GG (1979) Extracellular potassium in the mammalian central nervous system. Ann Rev Physiol 41:159-177
10. Treherne JE, Schofield PK (1979) Ionic homeostasis of the brain microenvironment in insects. Trends in Neurosci 2:227-230




Acknowledgements

This work was supported by the Deutsche Forschungsgemeinschaft




[*]Abteilung Biologie, Ruhr-Universität, 4630 Bochum / FRG

[+]Fakultät für Biologie, Universität Konstanz, 7750 Konstanz / FRG



Discussion

ten Bruggencate: 1) Do Retzius cells have action potentials with after hyperpolarizations with high $K^+$-conductance ($G_K$) that one could use to get an AHP-reversal potential which then could be compared with the $K^+$-equilibrium potential? 2) Usually the Na pump hyperpolarizes cells. Why do you get, despite this, such low membrane potentials?
Deitmer: 1) This would certainly be an interesting comparison. The action potentials of Retzius cells, which passively invade the somata, do have a small AHP. We have, however, not done these experiments. 2) There might be several reasons for the relatively low membrane potential of Retzius cells; (1) these cells spontaneously fire action potentials, therefore it is difficult to give values for a "resting" membrane potential; (2) the cell membrane certainly has other ionic permeabilities besides that to $K^+$ ions which might be quite significant; and (3) the Na pump might contribute very little if at all to the membrane potential at normal steady-state conditions; we have indeed found that the Na pump can contribute a large amount to the membrane potential (>10 mV), when the Na pump is re-activated after inhibition in $OmK^+$ (Deitmer & Schlue, J Exp Biol, in press). Moreover, the values given here for the membrane potential of Retzius cells are very similar to those described by several other authors, and which we have also found using single-barrelled microelectrodes with very sharp tips.

# The Role of Extracellular Potassium and Hydrogen Activities in the Brain Cortex for Regulation of Cerebral Microcirculation in the Cat During Generalized Seizures and Specific Sensory Stimulation

E. LENIGER-FOLLERT, C. DANZ

It is well established that activation of the brain is accompanied by both increased cerebral metabolism and increased cerebral regional blood flow and that a tight coupling exists between functional activity and regional blood flow (15). In our studies we have shown that this tight coupling of cerebral flow to functional activity exists also in the microcirculatory range during the whole course of seizures and during specific sensory stimulation of the somatomotor cortex (7,8,9,11). Fig. 1 shows an original recording of arterial blood pressure, endtidal $CO_2$ content, electrocorticogram, cerebral microflow and local tissue $Po_2$ during bicuculline induced seizures. A decrease in $pH_2$ means an increase in microflow. It can be seen that microflow increases when the first changes in ECoG are observed. Maximum hyperemia occurs in parallel with maximum total power of the ECoG as analysed by Fourier transformation. During the period of oscillating functional activity microflow oscillates in parallel to the changes in ECoG. It decreases during the silent phases and increases during the epileptic discharges. When the period of rhythmical discharges and intermediate silent phases ceases microflow continuously decreases within a few minutes to a new steady-state which lasts until the end of experiments.

In experiments performed together with K.-A. Hossmann (8) we found that microflow increased in the somatomotor cortex of the cat during specific sensory activation produced by stimulating the contralateral forepaw with electric pulses of 0.3 ms duration and varying amplitude and frequency. The increase in microflow was limited to an area of 1 to 2 mm in diameter and was stronger the stronger the amplitude of the primary evoked potentials. Within 2 s after the onset of stimulation of the contralateral forepaw microflow increased in the right somatomotor cortex. When stimulation ended microflow returned to its initial level. When the ipsilateral forepaw was stimulated no change in microflow was seen in the right somatomotor cortex.

The question arises which factors are involved in this effective coupling of function and flow. From microapplication studies performed with artificial liquor in the perivascular space by Wahl et al (16,17), Kuschinsky et al (5), Betz et al (2,3), Berne et al (1) and Rubio et al. (13) it is known and generally accepted that several local factors are involved in the regulation of flow and that they interact with each other, for example extracellular $H^+$, $K^+$, and $Ca^{2+}$ activities, adenosine content, various transmitter substances and a possible hypoxia (18). In recent studies we have been able to demonstrate that hypoxia does not occur in the tissue during activation and that blockade of the ß-adrenergic and cholinergic nerves does not change the principal behaviour of microflow in the activated tissue (10, 11). Therefore, the aim of our present studies was to examine the time course of changes in extracellular $K^+$ and $H^+$ activities during the whole period of bicuculline induced seizures as well as during the specific sensory stimulation in order to see whether these two factors contribute to the flow regulation.

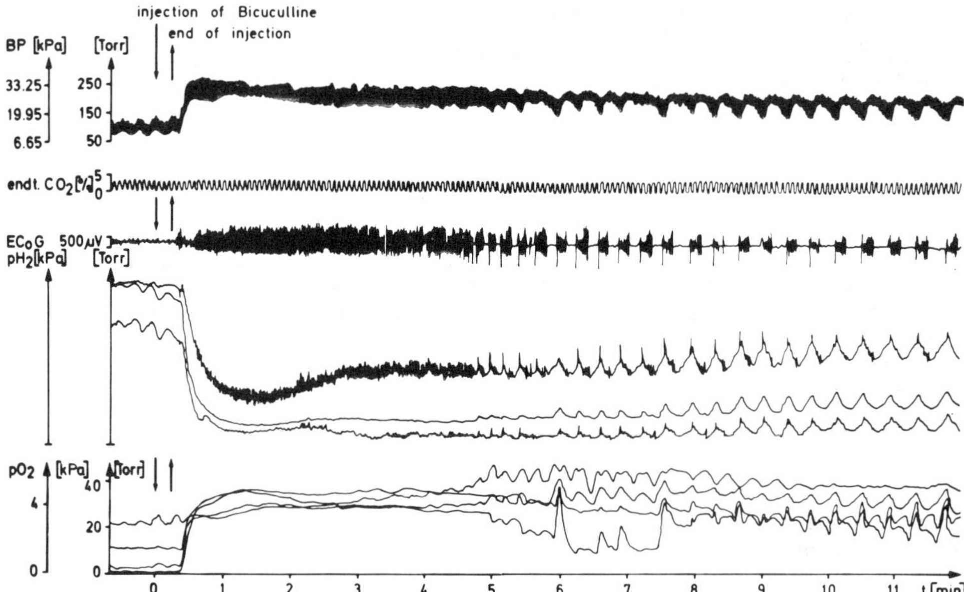



Fig. 1. Original simultaneous record of arterial blood pressure, endtidal $CO_2$ content, electrocorticogram, cerebral microflow and local tissue $Po_2$ before and during bicuculline induced seizures. A decrease in $pH_2$ means an increase in microflow

Experiments were performed in adult cats, anaesthetized with Nembutal (25-30 mg/kg body mass) or Chloralose(70 mg/kg), immobilized with Flaxedil and ventilated artificially with normal air. Arterial blood pressure, body temperature, endtidal $CO_2$ content and the acid base status of the blood were controlled and were in the normal range.
Extracellular $K^+$ activity was recorded either with double barrelled ion-sensitive microelectrodes with tip diameters of 1 to 4/um according to Lux and Neher (12) or with a solid contacted surface electrode according to Kessler,Höper and Simon (4).
Extracellular $H^+$ activity was monitored with $H^+$ sensitive glass micro-electrodes, filled with buffer solution according to Saito, Baumgärtl and Lübbers (14).The $H^+$ sensitive tip was 1 to 3/um in diameter and about 1 to 30/um long. We only used electrodes which showed a response time for 90% of 1 s or less and a reaction time of milliseconds.
This rapid response time was obtained by keeping the electrodes in dis-tilled water for about 3 to 5 months. In all experiments we used an Ingold Ag/AgCl-electrode via an agar bridge as an external reference electrode positioned on the surface of the brain about 1 to 2 mm from the inserting point of the measuring electrode. In some experiments we used a micro-pipette filled with 150 mM/l NaCl as a second external reference electrode that was located at the same depth as close as possible to the measuring electrode. Thus, it was possible to correct the $H^+$ signal for DC-poten-tials. To exclude mechanical artefacts it was necessary to mount the electrodes on a counterbalanced holder so that the electrodes could follow the movements of the brain.

102

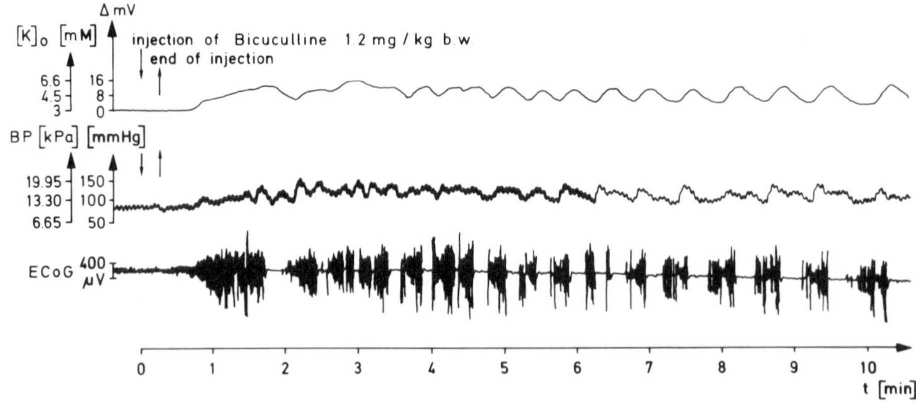



Fig. 2. Original recording of extracellular K$^+$ activity during bicuculline
induced seizures with a solid contacted valinomycin surface electrode. Note
the oscillations of extracellular K$^+$ activity in parallel to the functional
activity in ECoG

## Results and Discussion

During bicuculline induced seizures we obtained the following results:
1) K$^+$ activity was measured in 7 cats. In all experiments local extracellular K$^+$ activity immediately increased with the onset of epileptic discharges as shown in Fig. 2. It shows an original recording of K$^+$ activity
corrected for DC-changes with a valinomycin surface electrode. In parallel
to the repetitive silent and non-silent phases, extracellular K$^+$ activity
oscillated. It decreased when the flattening of ECoG began and increased
again when the discharges reappeared. When the seizures completely ceased
K$^+$ activity rapidly decreased to control levels or just below control
values (not included in the figure). We obtained the same results when we
inserted the double barrelled microelectrodes (500 to 1500/um) into the
tissue.
2) Extracellular H$^+$ activity (recorded in 9 cats) increased in all experiments a few seconds after the onset of discharges, in most cases without an
alkaline shift. Fig. 3a shows as a typical example the kinetics of H$^+$ activity together with the changes in ECoG. Maximal increase occurs in parallel to maximal functional activity. H$^+$ activity also shows rhythmical oscillations during the period of repetitive silent phases and bursts. It decreases during the silent phases and re-increases after the onset of new
bursts. Fig. 3b shows the kinetics of these oscillations in detail. When
the seizures are finished H$^+$ activity decreases to control level within a
few minutes.

When the right somatomotor cortex was specifically stimulated we obtained
the following results:
1) Local extracellular H$^+$ activity was measured together with evoked potentials in the activated somatomotor cortex in 4 cats. Amplitude, frequency and time of stimulation of the contralateral forepaw were varied. Whereas the behaviour of evoked potentials showed a stimulation dependent ampli-

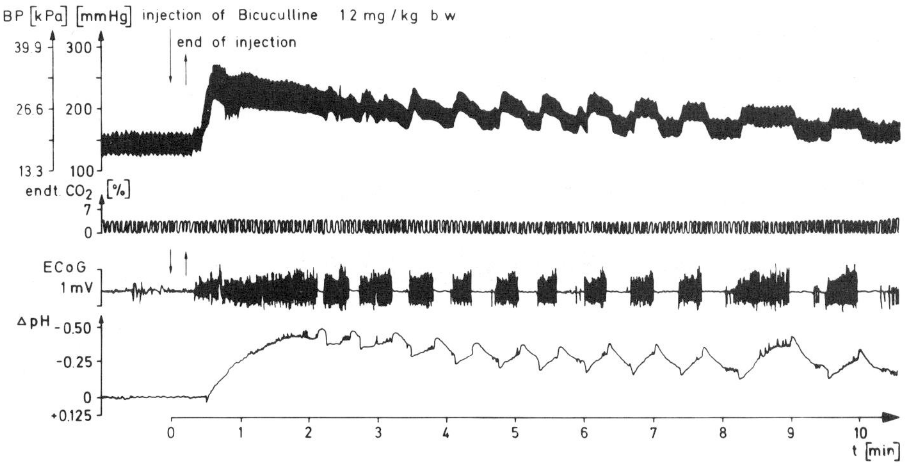



Fig. 3a. Original recording of arterial blood pressure, endtidal $CO_2$ content, ECoG and extracellular $H^+$ activity. Note the increase in $H^+$ activity which occurs some seconds after the onset of seizures

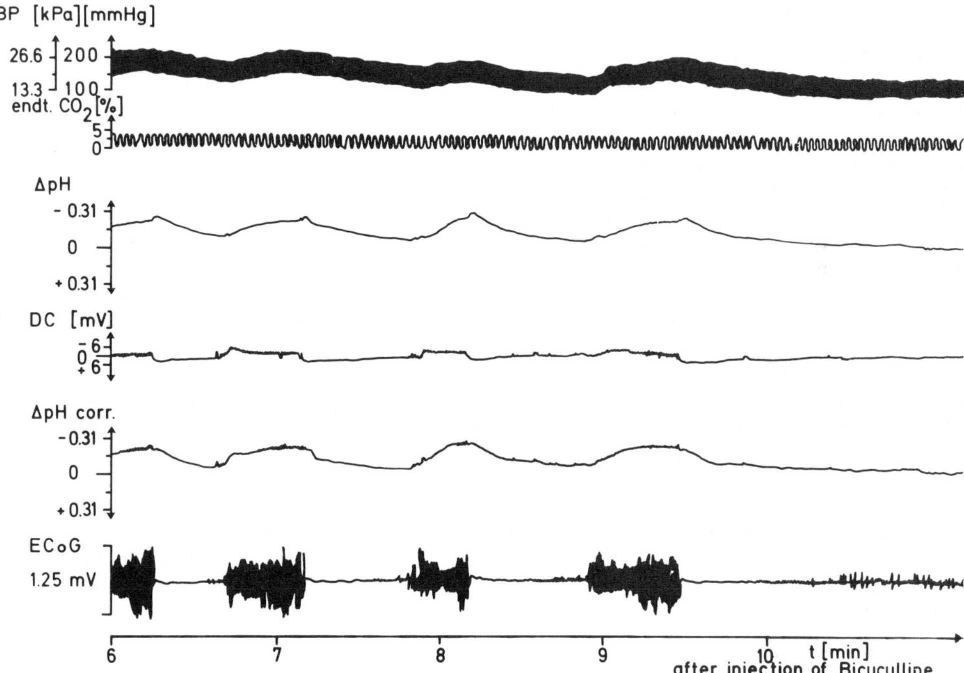



Fig. 3b. Simultaneous recording of arterial blood pressure, endtidal $CO_2$ content; uncorrected extracellular brain $H^+$ activity, DC-potential, corrected $H^+$ activity and ECoG. Note the oscillations of the corrected $H^+$ activity in parallel to the changes in ECoG in the later phase of seizure

tude as reported earlier (8), local extracellular $H^+$ activity did not change at all in the activated area where the increase in microflow occurred.

2) Local extracellular $K^+$ activity, however, increased immediately in the activated area after the onset of stimulation of the contralateral forepaw. It reached a maximum at the end of indirect stimulation. Then it rapidly decreased to its initial level with an undershoot.

We conclude from these results that extracellular $K^+$ activity may be one of the main mechanisms triggering and maintaining the flow increase during strong and very moderate activation of the tissue. Extracellular $H^+$ activity obviously contributes to the flow regulation only when activation is strong, such as during generalized seizures and during direct electrical stimulation of the brain tissue (6), when blood flow increases about 3 to 5 fold of the control values as demonstrated with the $C^{14}$-Jodo-antipyrine-method. In those cases the metabolism of the brain neurons increases so strongly that the generated $H^+$ ions cannot be completely buffered by the intracellular systems. However, during physiological activation, such as indirect stimulation of the brain cortex, when the flow increase is very moderate, maximally up to 60% (8), extracellular $H^+$ activity does not contribute to the flow increase.


## References


1. Berne RM, Rubio R, Curnish RR (1974) Release of adenosine from ischemic brain. Effect on cerebral vascular resistance and incorporation into cerebral adenosine nucleotide. Circulat Res 35: 262
2. Betz E, Enzenross HG, Vlahow V (1973) Interaction of $H^+$ and $Ca^{2+}$ in the regulation of local pial vascular resistance. Pfluegers Arch 343: 79
3. Betz E, Czornai M (1978) Action and interaction of perivascular $H^+$, $K^+$, and $Ca^{2+}$ on pial arteries. Pfluegers Arch 374: 67
4. Kessler M, Höper J, Simon W (1974) Methodology and application of multiple ion-selective surface electrode (pH, pK, pNa, pCa, pCl) for tissue measurements. Fed Proc 33: 279
5. Kuschinsky W, Wahl M, Bosse O, Thurau K (1972) Perivascular potassium and pH as determinants of local pial arterial diameter in cats. A microapplication study. Circulat Res 31: 240
6. Leniger-Follert E, Urbanics R, Lübbers DW (1978) Behavior of extracellular $H^+$ and $K^+$ activities during functional hyperemia of cerebral microcirculation. In: Cervós-Navarro J, Betz E (eds) Advances Neurol (20. Ed). Raven-Press, New York. pp 97-101
7. Leniger-Follert E (1979) Kinetics of microflow and extracellular $K^+$ activity in the brain cortex during bicuculline induced seizures. Proc Internat Soc Oxygen Transport to Tissue, July 27-29. La Jolla, Calif
8. Leniger-Follert E, Hossmann K-A (1979) Simultaneous measurement of microflow and evoked potentials in the somatomotor cortex of the cat brain during specific sensory activation. Pfluegers Arch 380: 85
9. Leniger-Follert E, Klasen KP (1980) Kinetics of microflow and local tissue $Po_2$ in the brain cortex (cat) during bicuculline induced seizures. Pfluegers Arch Suppl 384: R5

10. Leniger-Follert E (1980) Mechanisms of regulation of cerebral microcir-
    culation during bicuculline induced seizures. Proc Internat Union of
    Phys Sciences Vol XIV. Budapest. p 545
11. Leniger-Follert E (1980) Microflow and oxygen supply of the brain
    during local and generalized activation. Intern Soc on Oxygen
    Transport to Tissue. Budapest. in press
12. Lux HD, Neher E (1973) The equilibration time course of $K^+$ in cat
    cortex. Exp Brain Res 17: 190
13. Rubio R, Berne M, Winn HR (1978) Production, metabolism and possible
    functions of adenosine in brain tissue in situ. In: Cerebral vascu-
    lar smooth muscle and its control. Ciba Foundation Symp 56, Excerpta
    Medica. Elsevier, Amsterdam-Oxford-New York. pp 355-373
14. Saito Y, Baumgärtl H, Lübbers DW (1976) The RF-sputtering technique as
    a method for manufacturing  needle-shaped pH microelectrode. In:
    Kessler M, Clark LC jr, Lübbers DW, Silver IA, Simon W (eds) Ion and
    enzyme electrode in biology and medicine. Urban & Schwarzenberg,
    München-Berlin-Wien. pp 103-109
15. Sokoloff L (1978) Local cerebral energy metabolism: its relationship
    to local functional activity and blood flow. In: Cerebral vascular
    smooth muscle and its control. Ciba Foundation Symp 56. Excerpta
    Medica. Elsevier, Amsterdam-Oxford-New York. pp 171-191
16. Wahl M, Deetjen P, Thurau K, Ingvar DH, Lassen NA (1970) Micropuncture
    evaluation of the importance of perivascular pH for the arteriolar
    diameter on the brain surface. Pfluegers Arch 316: 152
17. Wahl M, Kuschinsky W (1976) The dilatatory action of adenosine on pial
    arteries of cats and its inhibition by theophylline. Pfluegers Arch
    362: 55
18. Wei EP, Kontos HA, Rosenblum WJ, Patterson JL jr (1977) Reduction of
    seizure-induced cerebral vasodilatation by increased local delivery
    of oxygen. in: Ingvar DH, Lassen NA (eds) Cerebral function, metabo-
    lism and circulation. Munksgaard, Copenhagen. pp 12.6-12.7


## Acknowledgements

The authors are indebted to Prof. Dr. M. Kessler, Erlangen for the valinomycin surface electrode and Prof. Dr. D.W. Lübbers and Chem. Ing. (grad).H.
Baumgärtl, Dortmund for the $H^+$ sensitive microelectrode.


Max-Planck-Institut für Systemphysiologie, Rheinlanddamm 201, 4600 Dortmund
1 / FRG

# Possible Potassium Sources in Activated Loci of Mammalian Neocortex as Reflected by Electrical Currents and Potassium Changes in the Extracellular Space

J. MACHEK, E. UJEC, O. KELLER, V. PAVLIK

For the purpose of the present paper potassium ($K^+$) sources and sinks are structures or events which can be inferred to produce measurable $K^+$ increases or decreases, respectively, in the extracellular (EC) space.

In the giant axon of <u>Loligo</u>, the second half of the time course of the spike has been shown to coincide with an increase of $K^+$ permeability (3) causing $K^+$ to escape from the fibre and to accumulate in the extracellular space where its concentration is proportional to decreases in the positive phases of the action potentials (2).

Also in the mammalian brain EC, $K^+$ responses have been reported during spontaneous (10), or evoked action potential activity (1). On the other hand, interhemispheric stimulation which reliably enhances $K^+$ levels in the neocortex of rats rarely elicited the unitary action potential activity in our experiments (4, 5). Thus, also in accord with data reviewed recently (9), other events than action potentials have to be envisaged as sources of the excess EC $K^+$. In the present paper, relations between simultaneously recorded, evoked, phasic field potentials and $K^+$ changes will be used to outline the problems of determining the presumed $K^+$ sources and sinks in the mammalian neocortex which were activated by afferent somatosensory or interhemispheric stimulations.

## Methods

In seven cats and 80 rats chloralose and halothane was applied respectively during surgery which consisted of trephine openings over the primary somatosensory regions and drainage of the cisterna magna. $K^+$ sensitive, double-barrel microelectrodes or coaxial double-barrel ones were used with $10^{14}\,\Omega$ input impedance, negative capacity compensated, DC preamplifiers (11, 12). An Ag/AgCl reference electrode was placed in the neck muscles of the artificially ventilated rats and cats which were immobilized with curare and flaxedil, respectively. The concurrent extracellular $K^+$ responses and field potentials from the primary somatosensory areas were FM-tape recorded (DC-2500 Hz) or photographed from the oscilloscope (DC-10000 Hz). In the cats, electrical stimulation was applied to infraorbital or radial nerves (rectangular pulses, 0.03 - 0.1 ms, less than 15 V constant voltage, 0.1 - 3 Hz); in the rats, to the homotopic sensorimotor area contralateral to the recording site (rectangular pulses, 1 - 2 ms, 0.8 - 3 mA constant current, 0.3 - 100 Hz, train duration 60 - 180 s).

## Peripheral Nerve Stimulation

Several issues were illustrated by the experiments in which rapid, low noise, potassium selective, coaxial, double-barrel microelectrodes were used for recording in cats (11, 12).

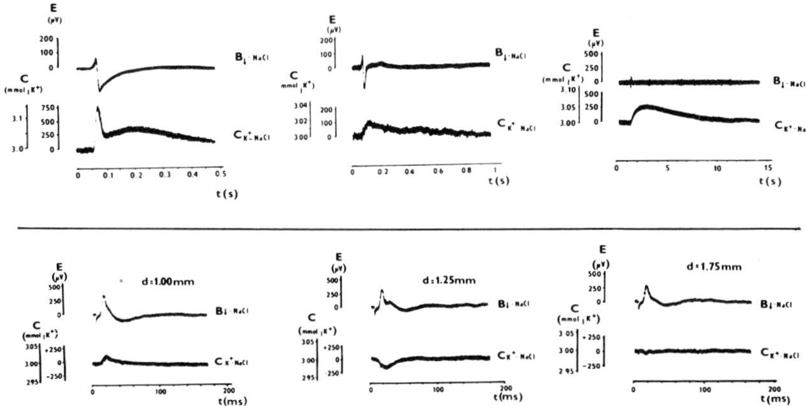

Fig. 1. Miniature $K^+$ transient and phasic field potentials. $B_{I-NaCl}$ - phasic field potential; $C_{K^+-NaCl}$ - $K^+$ transients; d - depth underneath cortical surface. Single stimuli to the radial nerve. One cat, two electrodes. Upper part - three separate penetrations. Lower part - one penetration at different depths

1. Miniature $K^+$ changes of distinctly different time courses were recorded in response to single stimuli which were applied at 10 s intervals to the radial nerve contralateral to the cortical recording area (Fig. 1, second row from the top). When graphically processed and transferred to a common time and voltage scale (Fig. 2), the miniature $K^+$ changes may be well compared and categorized.

Type-one miniature $K^+$ transients had a duration in the order of magnitude of tens of ms (40 - 50 ms in Fig. 1, top left; Fig. 2, a, b), a very steep rise (inset Fig. 2, a, b) and very short time constants in the order of ms (numerically in inset Fig. 2, a (4 ms), b (6 ms)).

Type-two miniature $K^+$ changes (Fig. 1, middle top; Fig. 2, c, d) lasted hundreds of ms, revealed a very steep or just steep rise (Fig. 2, inset c, d) and time constants in the order of tens of ms (Fig. 2, inset, see numerical values).

Type-three miniature $K^+$ responses lasted seconds (Fig. 1, top right; Fig. 2, e), their rises were flat and had time constants in the order of hundreds of ms.

2. The fastest miniature $K^+$ transients of the type-one were confined to small recording sites as illustrated in the bottom traces of Fig. 1. A displacement of the electrode tip by 250 um caused a $K^+$ increase (Fig. 1, bottom left) to transform into a decrease (Fig. 1, bottom middle). Five hundred um deeper a $K^+$ response was barely detectable (Fig. 1, bottom right). The EC phasic field potentials had the same polarity in all three recording sites. Localized $K^+$ decreases were observed also by Singer and Lux (8).

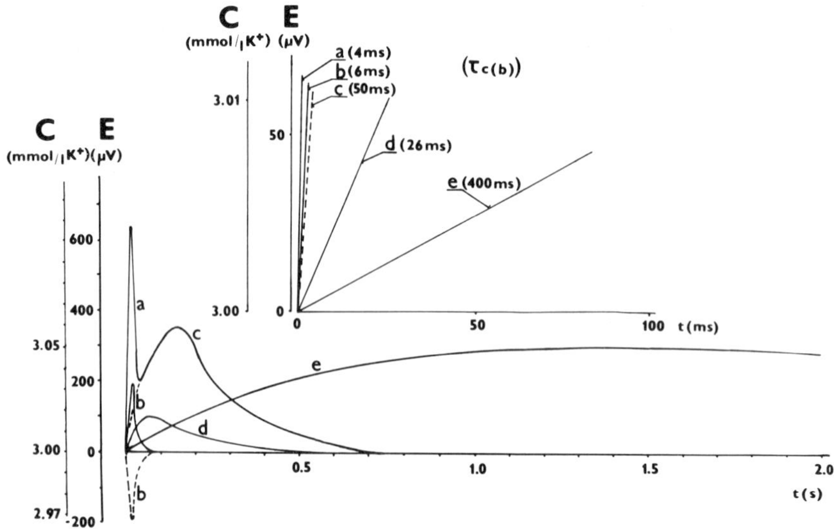

Fig. 2. Miniature $K^+$ transients of Fig. 1 transformed to common voltage and time scales. a-e - identification of curves; Insert: $\tau_{C(b)}$ - time constants (in brackets) and initial slopes of the curves a-e

3. EC unitary action potentials were very rarely recorded in coincidence with potassium responses, although the electrodes and the recording system were adequate for such purpose.

4. The phasic field potentials preceded the $K^+$ responses by several ms. The polarity of potentials concurrent with the type-one fast $K^+$ transients did not change with the polarity of the respective $K^+$ signals; however, differences in their time course were seen on decline (compare Fig. 1, bottom, the left and middle column traces). A negative after-potential coincided with the decay phase of the type-two $K^+$ change (Fig. 1, top, middle).

5. Summated $K^+$ responses coincided, of course, with negative slow potential changes during repetitive stimulation.


Interhemispheric Stimulation

The interhemispheric stimulation provides an intensive input of synchronized orthodromic and antidromic pulses into the cortical area homotopic to the stimulus site. Despite the input intensity, unitary activity was seldom observed even in experiments in which high $K^+$ activity levels of up to 12 mmol/l were recorded during 10 Hz repetitive stimulation (4, 5, 6) in curarized, artificially ventilated rats.

During ictal afterdischarges lasting tens of seconds the $K^+$ concentration was maintained approximately at the ceiling level of about 10 mmol/l with occa-

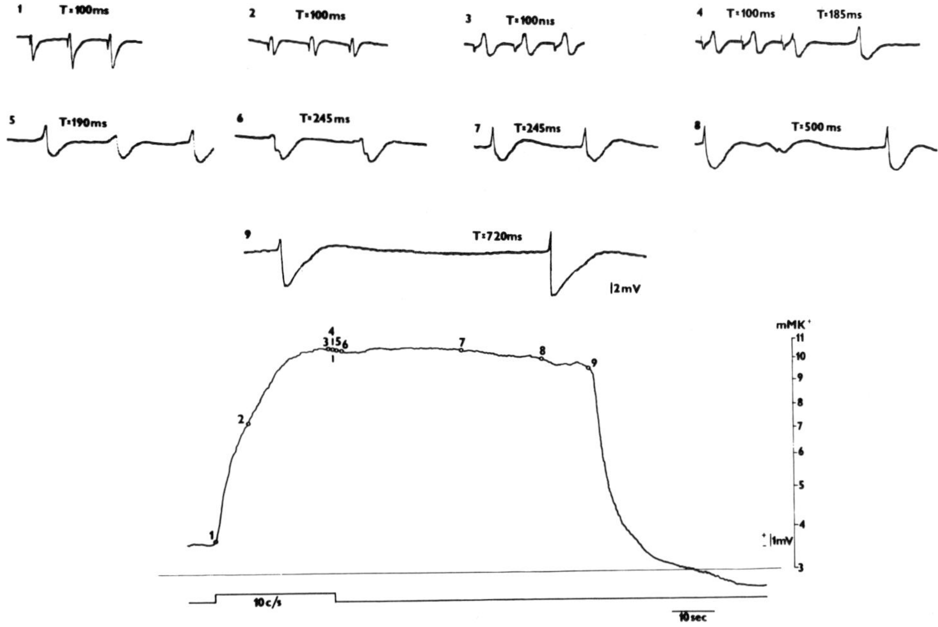

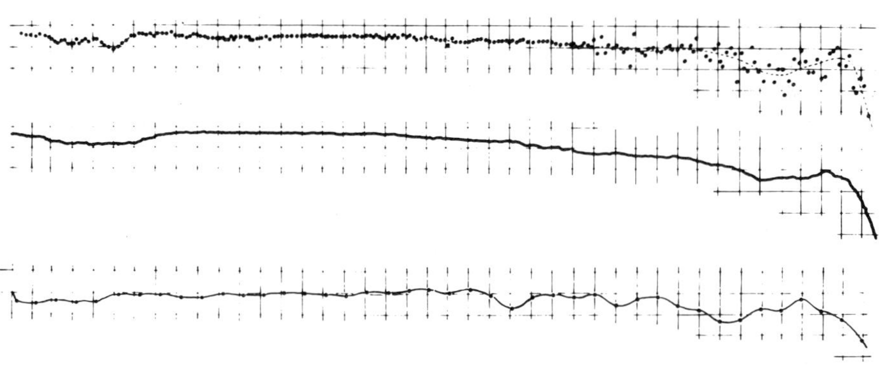

Fig. 3. Rhythmic interhemispheric stimulation 10 Hz, rat. Upper part - stimulus-locked and self-sustained field potentials. 1-9 - corresponding points of the $K^+$ curve in the middle part. Lower part - upper curve - frequency of afterdischarges; middle curve - $K^+$ concentration; lower curve - negative waves of the afterdischarges in arbitrary planimetric units. Abscisse-time calibration - 1.5 s/div for the lower part

sional decreases. These decreases concurred with phases in which the frequency of the afterdischarges also decreased and/or in which the negative components of the self-sustained activity became less ample and/or shorter. The graphical expression of this situation is in the lower part of Fig. 3. The oscillation of both the amplitudes and duration of the negative waves of afterdischarges are  characterized by the bottom curve, the ordinate of which is given in arbitrary surface units resulting from planimetric measurements. This curve is similar in time course to the $K^+$ concentration curve (Fig. 3, lower part, middle curve); so is the frequency curve with ordinate in Hz (Fig. 3, lower part, top curve). There was no similarity between the $K^+$ curve and the curves of other parameters of the self-sustained phasic activity, i.e. planimetric units of the positive components and mere amplitudes of the negative and positive waves (not shown). In the upper part of Fig. 3, phasic stimulus-locked field potentials and self-sustained afterdischarges are illustrated (note that no unit potentials are present) in samples labelled 1 - 9 which correspond to identical points of the $K^+$ concentration curve in the middle part of the figure.


Discussion

High resolving power and high fidelity of the recording system are essential for the classification of the miniature $K^+$ transients. The resolving power of the ISM, used in the experiments with the peripheral stimulation was substantially increased by shortening the column of the liquid ion exchanger (Corning 477317) and by using a smaller micropipette which was inserted into the ion sensitive channel (11). The fidelity of the records was tested by graphical processing of real recordings (12). To this effect three preamplifiers were used with a potassium selective, coaxial, double-barrel microelectrode; three potential time courses were simultaneously recorded: 1, a "monopolar $K^+$", i.e. $K^+$ sensitive barrel to ground; 2, "monopolar electric" i.e. the reference barrel to ground; and 3, the "$K^+$ differential" i.e. the $K^+$ sensitive to reference barrel. When the "$K^+$ differential" curve was graphically subtracted from the "monopolar $K^+$" curve, the "monopolar electric" curve was obtained, i.e. the extracellular field potential (12).

Peripheral nerve stimulation supplies central target areas where the $K^+$ changes were measured, with an input which is probably supramaximal to a physiological afferent impulse flow. Nevertheless the method is fine enough to resolve differences in time courses as well as in localization of $K^+$ sources and sinks in structural aggregates in the order of magnitude of tens of ms, $K^+$ umol/l and hundred micrometers. The slopes and time constants of the rise and the amplitudes of the $K^+$ transient characterize the potency of the sources. Since there is evidence of an active $K^+$ clearance in the brain (13), the decay parameters of the $K^+$ changes express the efficiency of the sinks. The distance of the electrode tip from the active region influences the amplitude and the initial slope of the $K^+$ signal but not so much the time constant. This is illustrated in Fig. 2: The curve "a" differs from the curve "b" by amplitude only; the broken curve "b" downwards, mirrors a $K^+$ sink displaying similar parameters as the "b" source (also Fig. 1, bottom middle); the curve "c" and "d" display time constants of the same order with various amplitudes and slopes. (They were classified on the basis of the time constants as type two). The time constants of half mmolar $K^+$ changes, obtained by Roitbak et al. (to be published), using direct cortical stimuli in cats under deep nembutal anaesthesia with 45 ms and 56 ms time constants (7, Fig. 1,2) belong in the type-two group.

The interhemispheric stimulation is used in the present report to corroborate one of the points evidenced previously, namely, the fact that important $K^+$ increases may occur without obvious participation of unitary action potentials as $K^+$ sources. The question of whether this absence of unit firing is due to some microelectrode bias or to some other artifact is not vital for the argument. Nevertheless, to eliminate the objection that a dead space around a double-barrel microelectrode tip may worsen the recording conditions for units we used single NaCl filled microelectrodes which (under stimulation known to increase $K^+$ up to 12 mmol/l) responded to the slow $K^+$ increase by a negative slow potential change (6). Important negative slow potential shifts occurred in the presence as well as in the absence of unitary bursts. Conclusive cases were those in which a stimulus locked unit appeared following an important focal depolarization, thus demonstrating the EC action potential not to be the cause of excess $K^+$ in the EC space but the consequence of a $K^+$ dependent focal depolarization.

## Conclusions

1. Using rapid, low-noise, potassium-sensitive, coaxial, double-barrel microelectrodes we differentiated three types of miniature $K^+$ changes the time constants of which differed by one order of magnitude (1 ms, 10 ms, 100 ms) in response to single stimuli of the radial and infraorbital nerves.

2. We presume that the different time courses of the $K^+$ transients correspond to distinct types of membranal events operating at specialized membranal or cellular aggregates.

3. The most rapid $K^+$ transients appeared to be sharply localized within hundred micron limits.

4. Rapid transient $K^+$ decreases were also recorded.

5. There was no correlation between the polarity of the evoked field potential and the polarity of the miniature $K^+$ signals.

6. Unit activity coincided exceptionally with miniature $K^+$ signals. Neither was its presence mandatory for important elevation of $K^+$ concentration elicited by repetitive interhemispheric stimulation.

7. During self-sustained afterdischarges concurring with elevated $K^+$ levels of about 10 mmol/l following interhemispheric stimulation, similar oscillations of the $K^+$ concentration, frequency of afterdischarges and size of the negative wave of the afterdischarges expressed in arbitrary planimetric units were detected.

## References


1.  ten Bruggencate G, Nicholson C, Stöckle H (1976) Climbing fiber evoked potassium release in cat cerebellum. Pfluegers Arch 367:107-109
2.  Frankenhäuser B, Hodgkin AL (1956) The aftereffects of impulses in giant nerve fibres of Loligo. J Physiol 131:341-376
3.  Hodgkin AL, Huxley AF (1952) A quantitative description of membrane current and its application to conduction and excitation in nerve. J Physiol 117:500-544

4.  Machek J, Ujek E, Pavlík V (1975) Cortical extracellular potassium con-
        centration during development of the interhemispheric response into
        the selfsustained afterdischarge. Physiol Bohemoslov 24:41-46
5.  Machek J, Ujek E, Pavlík V (1976) Extracellular potassium concentration
        and focal electrical potentials elicited in the cerebral cortex of
        rat by interhemispheric stimulation. Neuroscience Letters 2:147-152
6.  Machek J, Ujek E, Pavlík V, Horák F (1977) Interhemispheric field po-
        tentials, spreading depression and kindling. Neuroscience Letters 4:
        337-341
7.  Roitbak AI, Machek J, Pavlík V, Bobrov AV, Otcherashvili IV (to be pub-
        lished) Direct cortical response and extracellular potassium. In:
        Hnik P, Syková E, Vyklický L (eds) Ion selective electrodes and their
        use in excitable tissue.
8.  Singer W, Lux HD (1975) Extracellular potassium gradients and visual re-
        ceptive fields in the cat striate cortex. Brain Res 96:378-393
9.  Somjem GG (1979) Extracellular potassium in the mammalian central nervous
        system. Ann Rev Physiol 41:159-177
10. Syková E, Rothenberg S, Krekule I (1974) Changes of extracellular con-
        centration during spontaneous activity in the mesencephalic reticular
        formation of the rat. Brain Res 79:333-337
11. Ujec E, Keller O, Machek J, Pavlík V (1979) Low impedance coaxial $K^+$
        selective microelectrodes. Pfluegers Arch 382:189-192
12. Ujec E, Keller O, Kriz N, Pavlík V, Machek J (to be published) Low-impe-
        dance, coaxial, ion-selective, double-barrel microelectrodes and their
        use in biological measurements. Bioelectrochemistry and bioenergetics.
13. Vern BA, Schuette WH, Thibault LE (1977) $[K^+]_o$ clearance in cortex: A
        new analytical model. J Neurophysiol 40:1015-1023

Institute of Physiology, Videnská 1083, 14220 Prague 4 / CSSR


Discussion

It was questioned by E. Frömter whether one could be sure that the miniature
$K^+$ changes measured were genuine.
Machek:  The belief that the miniature $K^+$ changes which we measured are genuine is based on systematically collected experience (11, 12, 18, 20, 21)
on the transfer properties of microelectrodes, ion-selective double-barrel
ISM and coaxial double-barrel microelectrodes (ISCM). We have found that in
$K^+$ISM and $K^+$ISCM, the capacitance crosstalk of both channels is almost measurable with capacitance compensated inputs for signals under 10 KHz, the
more so for events with time constants in the order of 1 ms. This was prac-
tically verified by applying simulated pulses to the electrode tips.
There is also evidence inherent in the present records: the polarity of the
evoked potentials which were recorded from the reference barrel against
ground, did not change with changing polarity for the $K^+$ signals which were
recorded from the sensitive barrel against the reference barrel at the same
time. And lastly, the finding of a "primary $K^+$ sink" in the somatosensory
cortex confirms a similar finding by Singer and Lux (8) in the visual cortex.
A question again related to the validity of the observation of a $K^+$ undershoot was asked and the lack of unitary activity was remarked upon:
ten Bruggencate:  I feel that I have to "defend" somehow the observation of
a "primary $K^+$ undershoot" that Dr. Machek described. That $K^+$ undershoots
without previous $K^+$ rises exist has been found by Lux and his coworkers in

various structures (see Lux HD (1975) In: Brain Work. Ingvar HD, Lassen NA (eds) Munksgaard Copenhagen, pp 172-181 and Nicholson CN (ed) (1980) Dynamics of the brain cell microenvironment. Neurosci Res Prog Bull 18:177-322) and careful checks indicated that they are probably not due to electrical artifacts. In addition Heinemann et al. (Heinemann K, Lux HD, Zander KJ (1978) In: Iontophoresis and transmitter mechanisms in the mammalian central nervous system. Ryall RW, Kelly JS (eds) Elsevier/North Holland Biomed Press Amsterdam, pp 419-428) described potassium decreases induced by local application of noradrenaline. They had a slow time course and were interpreted as due to an activation of the Na pumps by noradrenaline. Also, an inhibition of spontaneous discharges may contribute to the occurrence of primary undershoots.

Nicholson: In regard to the question of recording unit activity in cortical structures, it is worth noting that stimulation of the cerebellar surface stimulates many thousand parallel fibers, all of which generate action potentials but which, nevertheless, cannot be recorded as units with any microelectrodes, because of the small size of these axons. Again, the dentritic action potentials in the cerebellum are rarely recorded extracellularly with microelectrodes because their characteristics are unsuitable for recording extacellularly. Thus, only a small class of somatic or axonal extracellular action potentials are usually seen with microelectrodes.

Machek: We have also evidence (4, 5, 6) that the unit activity is indeed rare in response to interhemispheric stimulation using electrodes capable of recording them (Fig. 4). In agreement with our experiments, Asanuma and Oamoto (15) and Crowel (17) found that the commonest response of units to interhemispheric stimulation is suppression of the unit activity. Also the primary evoked response is reported as resulting from the summation of postsynaptic potential changes on synaptically driven neurons near the recording electrode (19). Especially inhibitory postsynaptic potentials might be a potent source of $K^+$ (14).

In a curarized rat, interhemispheric stimulation (duration 60 s, intervals 10 min, pulse parameters as in Fig. 3) $K^+$ levels of up to 10 - 12 mmpl/l are indicated by ink recorded negative slow potential changes (see Fig. 3 and references 4, 5, 6) in the second and third lines. Heavy numbers 4, 13, 14, 15 - the serial numbers of stimulation which begins at the first negative downward shift of each ink record. The first and fourth lines - fast sweep DC records photographed from the oscilloscope which correspond to the sharp negative maximum amplitude swing of the 10 Hz stimulation no. 4, and to the start of stimulation nos. 13-15. Short bars underneath the first line indicate phasic evoked potentials on which a single unit discharge was superimposed (the unit would probably not be seen in the printed figure). 4, 14 - 10 Hz stimulation; 13, 15 - 30 Hz stimulation. Negativity downwards; oscilloscope calibration 5 mV, 100 ms - right; ink recorder calibration 30s, 5 mV - middle of 4; valid also for 13, 14. Note that a unit appeared only in stimulation 4 secondary to a slow focal depolarization between the third and fourth phasic field potential. Then the first unit appeared on the top of the initial positive wave of the fifth field potential. In the 30 Hz stimulations nos. 13 and 15 there were no apparent units, only rhythmic field potential oscillations with fading amplitudes were recorded. Consider also the shape of phasic field potentials during the 10 Hz stimulations. Following the negative DC shift, in stimulation 4 but not so much in stimulation 14, the initial positive waves of the physic interhemispheric response became more ample which indicates that the focus was depolarized by a dipole extrinsic to the dipole generating the phasic field potential (16). Similar records were interpreted as a possible glial contribution to focal depolarization (5). In stimulation 14 only decreasing negative down strokes coincide

with decreasing slope of the negative slow potential. This phenomenon also
fits into the Bishop and O'Leary pattern (16).

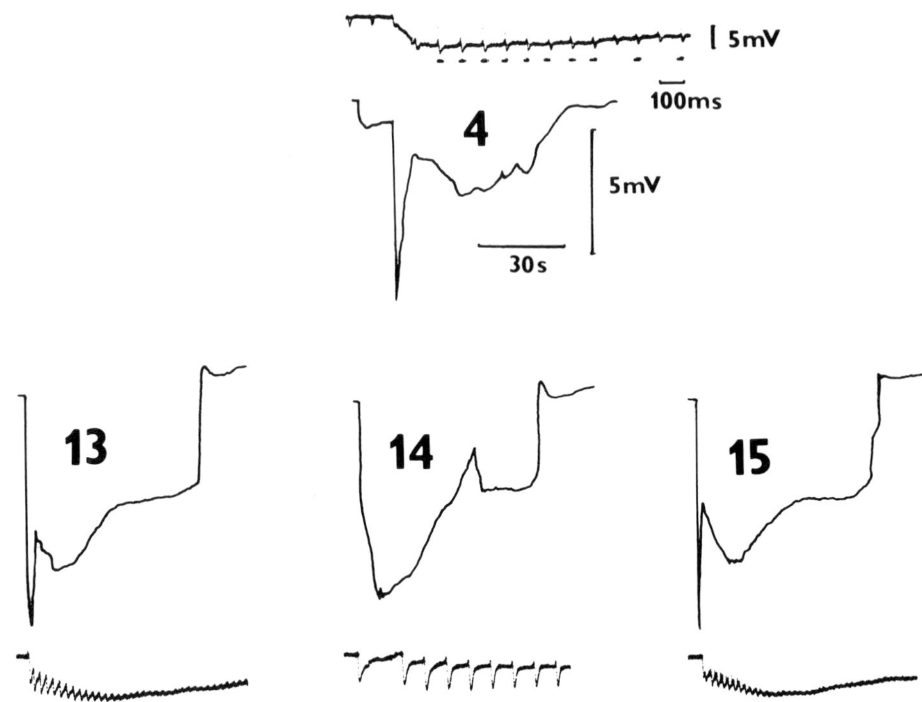


Fig. 4.  Unit activity in a less traumatic extracellular single microelectrode recording


A question from G.ten Bruggencate referred to the possible location of $K^+$
sources and sinks.
Machek:  It is pure speculation indeed when one tries to guess structural
$K^+$ sources and sinks from the parameters of the $K^+$ changes which are dependent on the unknown distance between the electrode tip and the presumed
source or sink. But the categorization of data is based on real findings
under standard conditions. Recording sites have been determined which may
be defined as $K^+$ sources, $K^+$ sinks and also $K^+$ neutral locations. They form
a mosaic within the activated focus and the size of the "grains" of the
mosaic is in the order of magnitude of hundreds of micrometers. I presume
also, that on the basis of our results as well as of those of Roitbak (7)
which were obtained under conditions which certainly suppressed unit activity,
one can draw a line between "action potential born $K^+$" and EC $K^+$ originating
from membraneal events different from the action potential. I think that the
contribution of the "extra action potential events" to the liberation of $K^+$
into the EC space is very important under our conditions and may be different
in other structures and other conditions.
There is more evidence available to the effect that action potential activity
may relatively little contribute to the EC $K^+$ levels resulting from stimulation experiments in the mammalian central nervous system. In rat mesence-

phalic reticular formation, spontaneous single unit potentials coincided with only small $K^+$ signals of O.O1 - O.O2 $mmol.l^{-1}$ estimated by computer averaging (10). On the other hand, climbing fibre evoked, i.e. Purkynie cell impulse activity dependent, $K^+$ increases in cat cerebellum amounted to O.3 - O.5 $mmol.l^{-1}$ per impulse (1). The comparison of both results (1 versus 10) indicates other events than initial segment limited action potentials as sources of EC $K^+$. Invasion of the dendritic tree by the action potential or synaptic currents are the possible mechanisms (9). No $K^+$ response in the ventral horn of the spinal cord following ventral root stimulation (23) corroborates the assumption of minor action potential contribution to excess EC $K^+$ elicited by electrical stimulation. The amount of $K^+$ liberated by an activated neuron also depends on the duration of the $K^+$ releasing events. This argument was used to explain the 5O times higher $K^+$ release resulting from a 1OO ms lasting photoreceptor depolarization as compared to small $K^+$ fluxes liberated by a 1 - 2 ms lasting action potential of the giant axon (22). Similar ratio exists between the duration of cortical inhibitory postsynaptic potentials and action potentials of the mammalian cortex; both events, besides other ones, are likely to release $K^+$ into the EC space (1, 9, 14).


Supplementary references from Dr. Machek's comments


14. Allen GI, Eccles JC, Nisoll RA, Oshima T, Rubia FJ (1977) The ionic
    mechanisms concerned in generating the IPSP of hippocampal pyramidal
    cells. Proc R Soc Lond /Biol/ 198:363-384
15. Asanuma H, Okamoto K (1959) Unitary study on evoked activity of callosal
    neurons and its effect on pyramidal tract cell activity in cats.
    Japan J Physiol 9:473-483
16. Bishop GH, O'Leary JL (1950) The effect of polarizing currents on cell
    potentials and their significance in the interpretation of central
    neurons activity. Electroenceph Clin Neurophysiol 2:401-416
17. Crowel RM (1970) Distant effects of a focal epileptogenic process. Brain
    Res 18:137-154
18. Rech F, Ujec E (1978) Transfer properties of the glass micropipette.
    Physiol Bohemoslov 27:81-89
19. Towe AL (1966) On the nature of the primary evoked response. Exp Neurol
    15:113-139
20. Ujec E, Keller O, Kriz N, Pavlík V, Machek J (to be published) The mea-
    surement of small and rapid ion concentration changes /$K^+$, $Ca^{++}$, $Cl^-$/
    using double-barrel, ion-selective, coaxial microelectrodes/ISCM.
    In: Hnik P, Syková E, Vyklický L (Eds) Ion selective electrodes and
    their use in excitable tissues.
21. Ujec E, Vít Z, Vyskocil F, Králík O (1973) Analysis of geometrical and
    electrical parameters of the tips of glass microelectrodes. Physiol
    Bohemoslov 22:329-337
22. Coles JA, Tsacopoulos M (1979) Potassium activity in photoreceptors,
    glial cells and extracellular space in the drone retina: changes
    during stimulation. J Physiol 290:525-549
23. Kriz N, Syková E, Ujec E, Vyklický L (1974) Changes of extracellular
    concentration induced by neuronal activity in the spinal cord of the
    cat. J Physiol 238:1-15

# The Kinetics of the Potassium Concentration in Brain Extracellular Fluid During Respiratory and Metabolic Disturbances

K. MÜCKENHOFF, H.R. AHMAD, A. LUTTMANN

## Introduction

It has been recently shown that the brain extracellular bicarbonate exchanges against chloride - similar to the Hamburger shift of the red cells - not only across the blood-brain-barrier but also with brain cells. The exchange ratio of bicarbonate and chloride was observed to be nearly one (1, 4).

So the questions to be posed are:
1. Can the existence of an independent $Na^+$- $K^+$-exchange be demonstrated across the blood-brain-barrier and glia cells?
2. Is there an influence of the brain extracellular $H^+$-ion activity on the mechanism of $K^+$-homeostasis?

The experiments presented here, therefore were planned in such a way that $[H^+]$, $[K^+]$ and $[Na^+]$ in the brain extracellular fluid could be measured continuously while metabolic or respiratory acid-base changes were induced in the blood.

## Material and Methods

Surface electrodes with flat membranes sensitive to $H^+$, $K^+$ and $Na^+$, along with a $CO_2$-electrode, were used in a type of application developed by Ahmad et al. (1). The pH-electrode was a conventional glass electrode. Also a conventional $CO_2$ electrode was used, consisting of a glass electrode and a $Ag^+$/AgCl reference, a silicon membrane and a bicarbonate buffer between membrane and electrode. The signal from this electrode was linearized and frequency compensated, yielding a 90 per cent response time of 1 s (5). For the determination of the extracellular potassium- and sodium-activities, ion-sensitive surface electrodes, consisting of a silver plate inserted into a small PVC cylinder and covered with sodium- and potassium-impregnated PVC membranes, were used (3, 7). These electrodes had 90 per cent reaction times of approximately 10 s. The simultaneous measurements of the $H^+$-, $K^+$- and $Na^+$-ion activities were performed with three highly stable amplifiers with impedance transformation and potential isolation.

The experiments were carried out on artificially ventilated cats anaesthetized with α-glucochloralose (40 mg/kg) and urethane (200 mg/kg). Body temperature and the temperature of the brain surface were measured and kept constant at $37.0^o$ C $\pm$ $0.2^o$ C. The pH, $pCO_2$-, $K^+$- and $Na^+$-electrodes were balanced on the brain surface (parietal cortex) in order to provide constant minimal force. The time courses of the $H^+$-, $K^+$-, and $Na^+$-ion activities in the brain extracellular fluid were observed either after intravenous injection of $K_2SO_4$ at constant $pCO_2$ or a step change of alveolar $pCO_2$. $CO_2$ was added by injection of $CO_2$ gas to the inspiratory side. Injections were given into the vena femoralis.

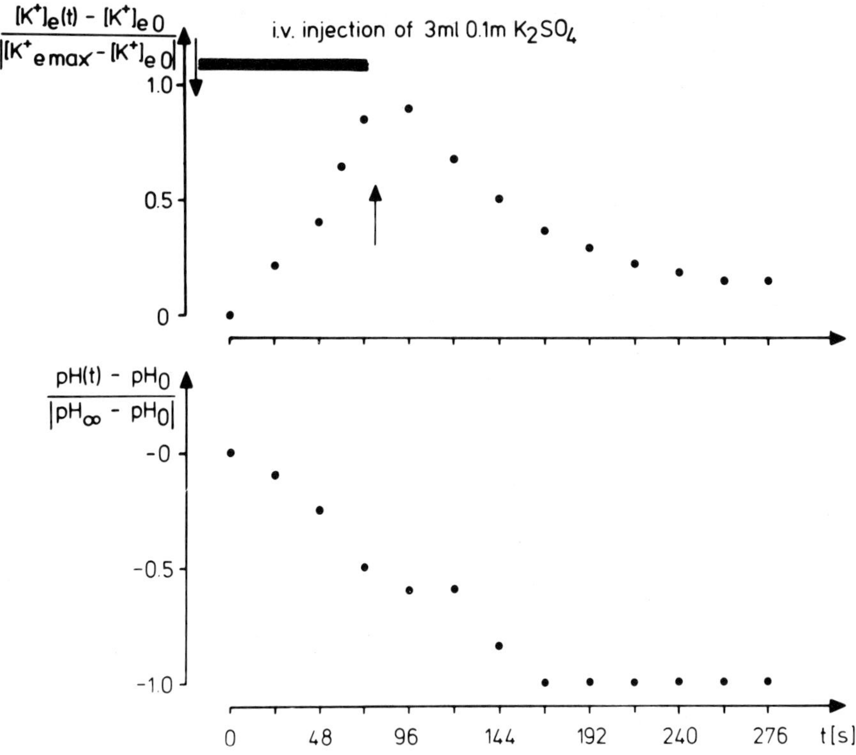



Fig. 1. Intravenous injection of 3 ml 0.1 m$K_2SO_4$ (= 0.1 mmol . $1^{-1}$) in an artificially ventilated cat causes a rise in the extracellular potassium activity. Extracellular $K^+$, pH and $pCO_2$ on the surface of the parietal cortex are recorded. From the original recording the pH- and $K^+$-signals are standardized and plotted against time. At a constant $pCO_2$ the maximal increase was: $\Delta[K^+]_e = +1.2$ mmol . $1^{-1}$ and $\Delta pH_e = -.03$


## Results

### Metabolic Influence

In Fig. 1 a typical experiment concerning the cation exchange at the blood brain barrier is shown where hydrogen- and potassium-ion activity have been recorded on the cortical surface of the brain of an artificially ventilated cat. It is assumed that the small liquid layer of the surface represents the extracellular fluid of the brain.

The potassium concentration of blood was raised by an intravenous injection of 3 ml 0.1 mmol . $1^{-1}$ $K_2SO_4$. After a delay of about 20 s the extracellular potassium concentration increases during injection up to a maximum of $\Delta[K^+]_e =$

$1.2 \text{ mmol} \cdot l^{-1}$. Soon after the end of injection an exponential decrease of
the extracellular potassium activity is observed. It does not return to the
control value completely. There remains a steady deviation of less than 0.1
$\text{mmol} \cdot l^{-1}$. The increase of the extracellular $H^+$-ion activity starts with the
increasing potassium-activity and continues within the phase of potassium return.

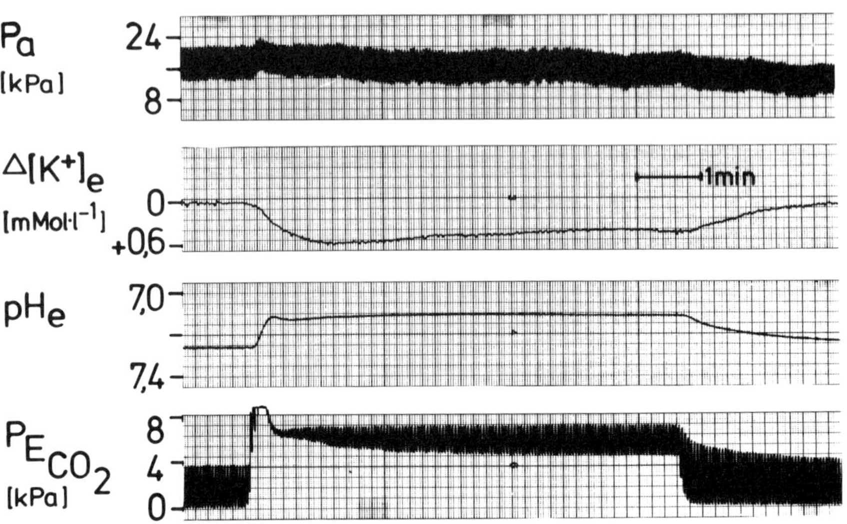



Fig. 2. During an approximately rectangular change of endtidal $pCO_2$ in an
artificially ventilated cat, arterial pressure, extracellular $[K^+]_e$ and pH
on the surface of the parietal cortex were measured

## Respiratory Influence

From a typical recording as shown in Fig. 2 the measured signals of extracellular hydrogen-, potassium- and sodium-ion activities were standardized
to one in an on-transient of a $P_A CO_2$ step and plotted against the time
(Fig. 3).

Fig. 3 demonstrates that an increase of hydrogen-ion activity in the extracellular fluid, and presumably also in the intracellular fluid, induces an
increase in the extracellular potassium and a decrease in the sodium activity.

## Discussion

It is concluded from both types of experiments that there exists a passive
potassium-sodium exchange at the blood-brain barrier and an active sodium-
potassium transport system across the glial membrane. This active transport
contributes to the $K^+$-homeostasis of the brain extracellular fluid. Increase
of $[K^+]_e$ leads to an increase of an active flux. In response to an intra-

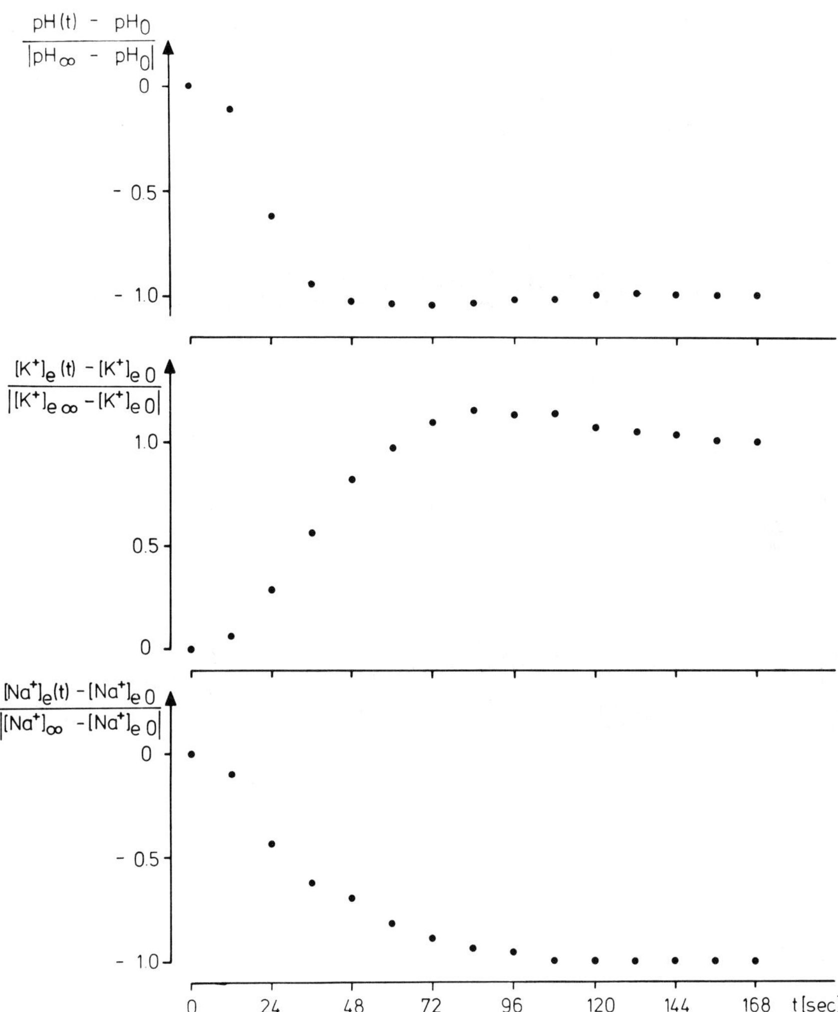

Fig. 3. From a recording of brain extracellular pH, $[K^+]_e$ and $[Na^+]_e$ are plotted in standardized form against time during the on-transient of an approximately rectangular step of $pCO_2$. There is a time delay of the potassium and sodium signals against pH, that may be partly due to the different time characteristics of the electrodes used, diffusion layer between the electrode and the brain surface and local circulation. The cat was artificially ventilated

venous $K_2SO_4$ injection $[K^+]_e$ increased transiently and returned almost to the control value while an increase in $[H^+]_e$ was observed.

It is known from experiments with human blood that changes of pH influence passive and active cation exchange: An increase of the extracellular and intracellular $H^+$ activity induces a decrease in the passive cation permeability which is in accordance with the concept of a fixed charged model of the membrane (6). On the basis of the tracer experiments which were performed on human red blood cells (2), which reveal a large decrease of the active sodium transport rate if the $H^+$ activity is increased, it may be stated from our experiments (Fig. 3) that acidosis causes a decrease of the active $Na^+$ $K^+$ transport which induces an increase in extracellular potassium activity and a decrease in sodium activity.

## References


1. Ahmad HR, Loeschcke HH, Woidtke HH (1978) Three compartments model for the bicarbonate exchange of the brain extracellular fluid with blood and cells. In: Fitzgerald RS, Gautier H, Lahiri S (eds) Regulation of respiration during sleep and anaesthesia. Plenum Press, New York, pp 195-209
2. Harris EJ, Mariels M (1951) The permeability of human erythrocytes to sodium. J Physiol 113:506-524
3. Höper J, Kessler M, Simon W (1976) Measurement with ion-selective surface electrodes (pK, pNa, pCa, pH) during no-flow anoxia. In: Kessler M, Clark LC Jr, Lübbers DW, Silver IA, Simon W (eds) Ion and enzyme electrodes in biology and medicine. Urban & Schwarzenberg, München Berlin Wien, pp 331-334
4. Loeschcke HH, Ahmad HR (1980) Transient and steady states of bicarbonate chloride relationship of the brain extracellular fluid. In: Bauer C, Gros G, Bartels H (eds) Biophysics and physiology of carbon-dioxide. Springer, Berlin Heidelberg New York, p 439
5. Luttmann A, Mückenhoff K, Loeschcke HH (1978) Fast measurement of $CO_2$ partial pressure in gases and fluids. Pfluegers Arch 375:279-288
6. Passow H (1969) Passive ion permeability of the erythrocyte membrane. Progr Biophys 19:423-467
7. Simon W, Pretsch E, Ammann D, Morf WE, Güggi M, Bissing R, Kessler M (1975) Recent development in the field of ion-selective electrodes. Pure Appl Chem 44:612-626


## Acknowledgements

We are grateful to Dr. W. Simon for providing a sodium exchanger.

Ruhr-Universität Bochum, Institut für Physiologie, Lehrstuhl I. Postfach 2148, 4630 Bochum-Querenburg / FRG

## Discussion

Kessler: Could you explain the meaning of your findings in terms of the philosophy, as it were, behind their occurrence?

Ahmad: I would like to make a comment on Dr. Kesslers' question about the "philosophy" behind the brain Hamburger shift and why cells are loosing $HCO_3^-$ when $CO_2$ is locally increased. In the brain extracellular fluid (ECF) the only important buffer is the bicarbonate ion. This ion cannot buffer against carbonic acid, because in the reaction of $H^+$ (from $H_2CO_3$) with $HCO_3^-$, as much

$HCO_3^-$ will be formed as disappears. Therefore, _in vitro_ an increment of $pCO_2$ is accompanied by a high increase of $H^+$ and almost no change of $HCO_3^-$. On the other hand, when brain ECF is examined _in vivo_ under different $CO_2$ loads, there is an immediate and marked increase in ECF bicarbonate with increase of $pCO_2$, yielding a typical $CO_2$ dissociation curve. The slope of this curve is similar to that of true plasma. This may well be explained by assuming that poorly buffered brain ECF, when in close contact with cells (glia), shares the intracellular buffering power by anion exchange processes in much the same pattern as does the plasma in contact with red cells. This hypothesis has been recently established by experimental evidence (Loeschcke and Ahmad 1980). These investigations led to a new concept regarding the exchange processes of $HCO_3^-$ and $Cl^-$ between blood plasma and the brain ECF on one hand, and between the cells of CNS (glia) and the extracellular fluid on the other hand. In respiratory acidosis the latter exchange is the leading process in such a way that $CO_2$ from blood enters the interstitial fluid and the brain cells without restriction, then it reacts with the intracellular buffers forming $HCO_3^-$ which in turn is exchanged against $Cl^-$. An exchange ratio of nearly one was observed because the brain ECF has no sink for $CO_2$. Similar rapid $HCO_3^-/Cl^-$ exchange has been observed at the blood-brain barrier during metabolic disturbances in the blood while keeping $pCO_2$ constant.
So we regard that the brain Hamburger shift plays a vital role in maintaining the acid-base homeostasis of the brain extracellular fluid and thereby the neural function.

# Role of Calcium in the Chemoreceptive Process of the Carotid Body

H. ACKER, M. DELPIANO, M. FISCHER, F. PIETRUSCHKA, R.G. O'REGAN

The chemoreceptor process in the carotid body can be regarded as being a $pO_2$ or $pCO_2$ dependent transmitter release from type-I cells exciting the nerve fibres which are synaptically connected to these cells. As known from other secretory cells, such as pancreas cells (10), the release of transmitter depends on the presence of extracellular calcium. Eyzaguirre and Zapata (3), Acker et al. (1) and Delpiano and Acker (2) have shown that the presence of extracellular calcium is essential for the chemoreceptor activity. The chemoreceptor becomes silent and unreactive to hypoxia or hypercapnia, when the calcium content of the perfusion or superfusion medium is diminished. In contrast to this Roumy and Leitner (9) suggest that under hypoxia the calcium uptake of the mitochondria is blocked, which induces an increase of the intracellular calcium level and, as a consequence, a transmitter release. The present paper describes attempts to substantiate the importance and necessity of extracellular calcium for the $pO_2$ and $pCO_2$ dependent transmitter release from the chemoreceptor cells in the carotid body.

The first figure shows that the extracellular calcium activity of the carotid body measured by ion-sensitive microelectrodes, has two types of response to hypoxia. When the blood pressure increases under hypoxia the calcium activity decreases, whereas the calcium activity increases when the blood pressure drops to low values under hypoxia. This dependence on blood pressure could not be detected under hypercapnia. Under these conditions the calcium activity always increases. Also under sympathetic stimulation we found an increase in extracellular calcium activity without any change in blood pressure (7). Since it is not possible to decide whether these changes in calcium activity are real concentration changes, or changes in the calcium binding capacity of protein, we determine the calcium concentration of the whole carotid body by means of atomic absorption spectroscopy. The calcium concentration of the carotid body varies under normoxia between 10 and 80 nM/glomus. Similar values have been detected in the arterial wall by Siegel (11). The calcium concentration of the carotid body is changed under hypoxia and hypercapnia. When comparing both carotid bodies in one cat, the calcium concentration increases by about 400% under hypoxia or up to 70% under hypercapnia. These values are significantly different from the normal variation of the calcium content under normoxia. The different calcium changes under hypoxia from a clear relationship with the fall in blood pressure. In figure two it is shown that with increasing fall of blood pressure under hypoxia, expressed as percentage change from the initial value, the calcium content increases linearly. The calcium content might reach values up to 140 nM/glomus.

The calcium content under hypercapnia increases up to 50 nM/glomus. According to the formula $a = \gamma \cdot C$ (a = ion activity, $\gamma$ = activity coefficient, C = ion concentration), we can assume that the increases in calcium content under hypoxia and hypercapnia are also detected by the ion-sensitive electrodes when assuming a constant activity coefficient. The blood pressure

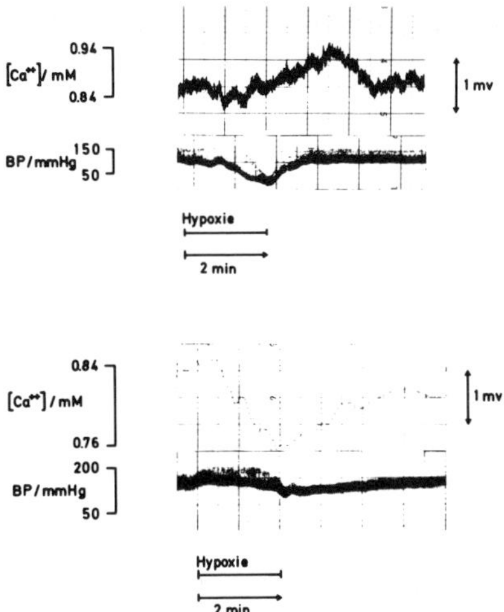



Fig. 1.  Increase and decrease in calcium activity in the cat carotid body woth changes in blood pressure during hypoxia

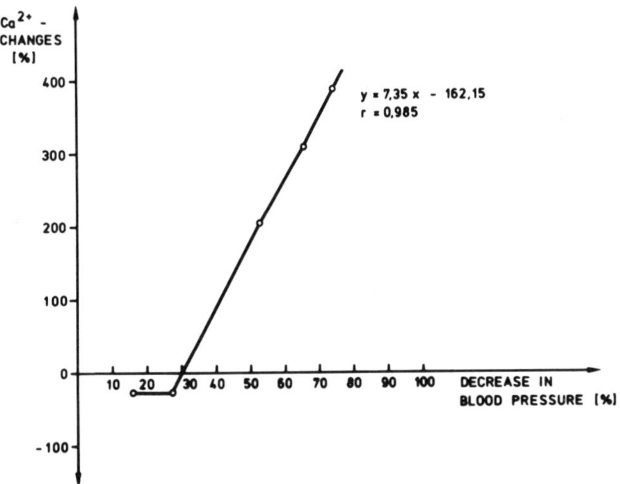



Fig. 2.  Dependence of the changes in calcium content on the decrease in blood pressure

dependent calcium changes under hypoxia could be caused by the relationship between changes in local and total flow. Under these conditions the local flow is generally decreased, but with higher total flow; i.e. with high blood pressure, the local flow is much more decreased. This means that, under hypoxic conditions with high blood pressure, the local flow is insufficient for the calcium supply in the carotid body. The calcium activity under hypoxia with high blood pressure, must be interpreted as a calcium flux into the cells to regulate transmitter release. Since the calcium supply is insufficient, the calcium extracellular activity must decrease. It seems that under hypoxic and low blood pressure conditions, the calcium influx into the cells is more than compensated by the local flow and the tissue becomes overfloated with calcium, measured as an increase in calcium content and calcium activity.

We investigated the question of the calcium influx into chemoreceptor cells on type-I cells in tissue culture, cultivated under normoxic and hypoxic conditions (8). In figure three, intracellular calcium activity and membrane

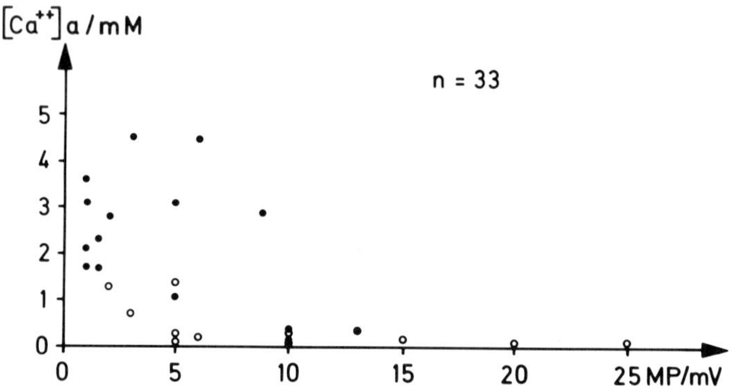


Fig. 3.  Intracellular calcium activity ($[\text{Ca}^{++}]_a$, y-axis) and membrane potential (MP, x-axis) of type-I cells cultured under normoxic (o) and hypoxic (·) conditions.

potential of chronically hypoxic (·) and normoxic (o) type-I cells are to be seen. Whereas normoxic cells have values for intracellular calcium activity between $10^{-5}$ and $10^{-3}$ M and membrane potentials up to -25 mV, the hypoxic cells have values of $10^{-3}$ M for intracellular calcium activity. These unusually high calcium levels are combined with low membrane potentials below -10 mV. Taking into account that Hansen (5) and Grönblad (4) found with electron microscopic studies that type-I cells can take up a large amount of calcium, especially the vesicles, our findings could mean that the type-I cells have taken up calcium activity by hypoxic stimulation. Pietruschka (8) could show that the mitochondria of these hypoxic cells have no detectable damage to ultrastructure. We therefore would like to exclude the mitochondria as a source of the high intracellular calcium. Since these cells have high intracellular potassium activities in spite of low membrane potentials, we suggest that type-I cells are able to concentrate intracellular calcium depending on $pO_2$ by an active process.

The increased, or decreased, extracellular calcium activity under hypoxia has a characteristic influence on the chemoreceptor discharge. This could be shown with the superfused carotid body in vitro. By changing the flow rate of the superfusion medium, the different extracellular calcium kinetics measured in vivo under hypoxia can be imitated. With a decreased calcium activity the chemoreceptor nervous discharge has a slow adaptation under hypoxia, whereas with increasing calcium activity, imitated with a low flow rate, the chemoreceptor discharge adapts rapidly.

Taking the present results into account, we would like to offer a model for chemoreception. Due to specific stimulation of the carotid body such as hypoxia, hypercapnia and sympathetic stimulation, as described by O'Regan (6), the calcium conductivity of the carotid body tissue is increased and calcium is taken up by the cells, for instance under hypercapnia. Depending on the local flow rate, which is constant or increased under hypoxia, hypercapnia and sympathetic stimulation, but more or less decreased under the second type of hypoxia, the calcium activity is increased (sufficient flow) or decreased (insufficient flow). These different levels of calcium activity induce different transmitter releases from the type-I cells with a dependence on the calcium threshold. Noradrenaline might be released from a low-threshold cell and dopamine from a high-threshold cell. Both transmitters induce different chemoreceptor responses: a more constant firing under hypoxia as induced predominantly by noradrenaline, or an adapting response under low blood pressure hypoxia, hypercapnia, and a depressed response under sympathetic stimulation, predominantly mediated by dopamine (which could be enlarged by the dopamine antagonist haloperidol (6)). Further experiments have to be done to prove the validity of this model.


## References


1.  Acker H, Weigelt H, Lübbers DW, Bingmann D, Caspers H (1976) Effect of changes in $Ca^{++}$ and $K^+$ activity upon tissue $pO_2$ and chemoreceptor activity of the cat carotid body. In: Paintal AS (ed) Morphology and mechanisms of chemoreceptors. Navchetan Press Ltd., New Delhi, pp 103-112
2.  Delpiano M, Acker H (1980) Extracellular $K^+$ and $Ca^{++}$ activities in the cat carotid body in vitro and their relationship to chemoreceptor activity during hypoxia and hypercapnia. Proc. I.U.P.S., Vol XIV, p 375
3.  Eyzaguirre C, Zapata O (1968) A discussion of possible transmitter or generator substances in the carotid body chemoreceptor. In: Torrance RW (ed) Arterial chemoreceptors, Blackwell, Oxford, pp 213-251
4.  Grönblad M, Åkerman KE, Eränkö O (1980) Ultrastructural evidence of exocytosis from glomus cells after incubation of adult rat carotid bodies in potassium-rich calcium-containing media. Brain Res. 189:576-581
5.  Hansen JT, Smith NKR (1979) Calcium binding sites in the vesicles of the carotid body and aortic body chief cells. Cell Tissue Res 199:145-151
6.  O'Regan RG (1977) Variable influences of the sympathetic nervous system upon carotid body chemoreceptor activity. In: Acker H, Fidone S, Pallot D, Eyzaguirre C, Lübbers DW, Torrance RW (eds) Chemoreception in the carotid body, Springer, Berlin Heidelberg New York, pp 160-167
7.  O'Regan RG, Fischer M, Acker H (1980) Extracellular $K^+$ and $Ca^{++}$ activity in the cat carotid body during sympathetic stimulation. Proc I.U.P.S., Vol XIV, p 623

8.  Pietruschka F, Schäfer D, Lübbers DW (1977) Reaction of cultured ca-
    rotid body cells to different concentrations of oxygen and carbon
    dioxide. In: Acker H, Fidone S, Pallot D, Eyzaguirre C, Lübbers DW,
    Torrance RW, Chemoreception in the carotid body, Springer, Berlin
    Heidelberg New York, pp 86-91
9.  Roumy M, Leitner LM (1977) Role of calcium ions in the mechanism of
    arterial chemoreceptor excitation. In: Acker H, Fidone S, Pallot D,
    Eyzaguirre C, Lübbers DW, Torrance RW, Chemoreception in the carotid
    body, Springer, Berlin Heidelberg New York, pp 257-263
10. Schulz I, Stolze HH (1980) The exocrine pancreas: The role of secre-
    tagogues, cyclic nucleotides, and calcium in enzyme secretion. Ann
    Rev Physiol 42:127-156
11. Siegel G, Jäger R, Nolte J, Bertsche O, Roedel H, Schröter R (1974)
    Ionic concentrations and membrane potential in cerebral and extra-
    cerebral arteries. In: Cervós-Navarro J, (ed),Pathology of cerebral
    microcirculation. Walter de Gruyter & Co., Berlin, pp 96-120


Max-Planck-Institut für Systemphysiologie, 4600 Dortmund / FRG


Discussion

Nicholson:  Since it has been suggested that high intracellular free $Ca^{2+}$
is likely to be cytotoxic, is it possible to make some estimate of the free
intracellular $Ca^{2+}$ level on the basis of your spectrophotometric measurement?
Acker:  No, unfortunately the carotid body is so small that the $Ca^{2+}$ cannot
be localized to any compartment, of which there are several.

# Intracellular Applications of $Ca^{2+}$-Selective Microelectrodes in Voltage-Clamped Shail Neurons

G. HOFMEIER, H.D. LUX

The calcium ion plays an outstanding physiological role besides the other intrinsic inorganic cations like sodium, potassium or magnesium, and this is especially true for the free intracellular calcium. Only an extremely small fraction of the total intracellular calcium is free in ionic form (3, 13). Estimates of $[Ca^{2+}]_i$ by various methods suggest that the resting level of intracellular calcium is at least 5 orders of magnitude lower than the value expected from a passive distribution (2, 9, 21). Neurons must possess especially powerful sequestration and pumping mechanisms for $Ca^{2+}$, since in excitable membranes inward calcium current densities in the order of $10^{-12}$ A/ $\mu m^2$ have been measured (15, 24). In absolute terms only a small amount of calcium is transferred into the cell by this current during electrical activity, but due to the low resting level, large relative changes in $[Ca^{2+}]_i$ occur. $Ca^{2+}$-dependent intracellular processes are thus strongly affected by the electrical activity of the neuron, with $Ca^{2+}$ acting as an intracellular messenger (16).

One such $Ca^{2+}$-dependent mechanism is the activation of a $K^+$-conductance $(g_K)$, which can be observed after influx or injection of $Ca^{2+}$ into the cell (19, 20). It has been claimed that this $g_K$ is proportional to $[Ca^{2+}]_i$ (20). We used $Ca^{2+}$-selective microelectrodes in combination with voltage-clamp and intracellular pressure injection to further investigate this question.

## Methods

The bursting pacemaker cell in the right parietal ganglion of the snail Helix pomatia (12) was used for the experiments, because $Ca^{2+}$-dependent $g_K$ plays an important role in the intrinsic discharge behaviour of this neuron (20). Due to its relatively large size of about 200 $\mu m$ diameter it is also well suited for the insertion of up to four intracellular micropipettes: two for the voltage-clamp, one $Ca^{2+}$-selective and one for pressure injection of $CaCl_2$ solution. $Ca^{2+}$-selective microelectrodes of either the double-barrelled or later the single-barrelled type were prepared following the methods described by Lux and Neher (18), using the liquid neutral-carrier ion-exchanger ETH 1001 (1). The injection experiments were done with electrodes employing the same sample of ion-exchanger which had already been used by Heinemann et al. (11) for extracellular work.

Extending the electrode calibration in that study to the intracellular range we calibrated at $Ca^{2+}$-concentrations from 10 mM down to 7 nM; all calibration solutions contained 100 mM KCl. The concentrations of less than 1 mM free $Ca^{2+}$ were prepared by the addition of 10 mM of calcium buffers. To cover the whole range of $Ca^{2+}$-concentrations we used nitrilotriacetate (NTA, Merck), N'-(2-hydroxyethyl)- ethylenediamine- N,N,N'- triacetate (HEDTA, Sigma) and ethyleneglycol- bis (oxyethylenenitrilo)- tetraacetate (EGTA, Merck). Since the apparent stability constants K' of these buffers are sensitive to variations in pH, this was adjusted with 10 mM of Tris/HCl. At pH 7.25 the values of K' calculated from tabulated data (5) are 1.18 x 10$^{-4}$ (NTA), 2.17 x 10$^{-6}$ (HEDTA) and 6.6 x 10$^{-8}$ (EGTA). Fig. 1A (a) shows the calibration curve obtained from these measurements. The responses were fitted to the modified Nernst-equation (14), using a nonlinear least-square computer fit. The resultant parameters were 29 mV/decade for the slope and a level of interference equivalent to 3.8 uM of calcium.

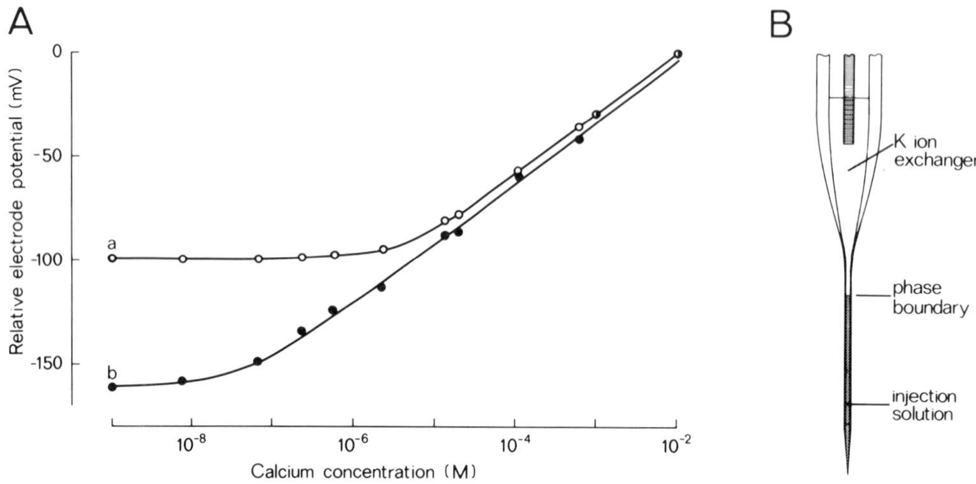



Fig. 1. A) Calibration curves for Ca$^{2+}$-selective microelectrodes made from first sample of ETH 1001 ion-exchanger (a) and freshly made from new sample (b). B) Construction of pressure pipette for quantifiable injections of aequous solutions

Single barrelled electrodes made more recently with a new sample of ETH 1001 ion-exchanger and tested with the same solutions gave a calibration curve as shown in Fig. 1A (b). These electrodes, when tested at the day of their preparation, display a limit of detection (14) of 30 nM of Ca$^{2+}$. This corresponds to a maximal potential difference between 10 mM Ca$^{2+}$ (the amount of Ca$^{2+}$ in normal snail ringer) and zero Ca$^{2+}$ (with 10 mM of EGTA added) of up to 160 mV. This limit of detection increases with time, however, and after between one and four days the calibration curve of these electrodes is very similar to the one shown in Fig. 1A (a). The source for this deterioration is probably the silanization of the glass tip, since the original detection limit of a deteriorated electrode is restored when it is freshly silanized and then refilled.

To test the response time of Ca$^{2+}$-selective electrodes to changes in $[Ca^{2+}]$ in the same range as those induced intracellularly in the injection experiments, HEDTA-buffered solutions were rapidly exchanged. The new readings were immediately stable after a period of artefacts due to the exchange procedure lasting for about 200 ms. Amplifiers with 10$^{15}$ Ohms input resistance and with compensated input capacitance were used throughout.

Measured quantities of 100 mM $CaCl_2$ solutions were injected with short (50 -
100 ms) pressure pulses through a pipette as shown in Fig. 1B, using a principle originally described by Llinas et al. (17). Instead of electrically
nonconducting oil as the hydrophobic phase in their design we used easily
available $K^+$ ion-exchanger (Corning 477317) and were thus able to check the
intracellular placement of the pipette by observation of its potential. The
pipettes were back-filled by an injection of the exchanger which spontaneously filled the tip. In a second step the injection solution was sucked up
through the tip. Application of about 10 nA of negative holding current was
found useful to prevent blockage of the tip when it was in the intracellular
milieu (7).

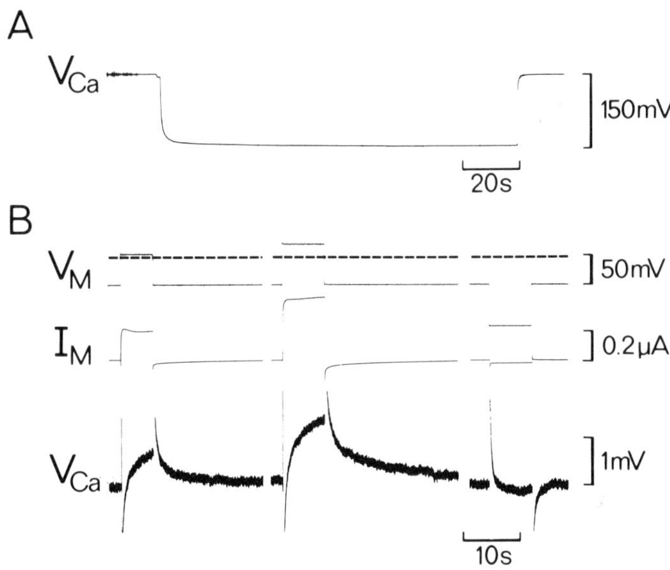


Fig. 2.   Responses of $Ca^{2+}$-selective microelectrodes with calibration curve
shown in Fig. 1Aa.   A) The electrode was inserted into a neuron voltage-
clamped at a constant holding potential of -50 mV and withdrawn again. B) Differential recording of $V_{Ca}$ showing rises in $[Ca^{2+}]$ due to voltage-dependent
calcium-influx. No calcium-signal with hyperpolarizing pulse


## Results

A typical trace of $V_{Ca}$ obtained on insertion and withdrawal of a $Ca^{2+}$-selec-
tive microelectrode into and out of a neuron voltage-clamped at a constant
holding-potential of -50 mV is shown in Fig. 2A. The electrode was of the
single-barrelled type, so that the membrane potential has to be subtracted
in this trace, and it had a limit of detection of 4 μM. On penetration of the
membrane $V_{Ca}$ fell very rapidly to nearly the final level with a half-time of
less than one second. However, there was a secondary slow phase, which took
one or several minutes to reach the last few mV. This is much longer than
the response time observed with changes from normal to low calcium-EGTA so-

130

lutions during the calibration procedure and is thus not produced by the electrodes. It seems more probable that intracellular $Ca^{2+}$-sequestration is disturbed around the penetrating electrode tip and that the time course of $V_{Ca}$ reflects the slow return of $[Ca^{2+}]_i$ from the calcium load associated with the penetration of the membrane.

From such measurements it could be concluded that $[Ca^{2+}]_i$ is lower than the detection limit of the electrodes, since the calibration curve is nearly horizontal at the intracellular values of $V_{Ca}$. Reliable measurements of the resting level were only possible with the more recent electrodes with a detection limit of 30 nM. While the time courses of $V_{Ca}$ on penetration were essentially the same as those shown for the less sensitive electrodes, these single-barrelled electrodes reached a mean intracellular $V_{Ca}$ of $-148 \pm 4$ mV (n = 5). The return to the baseline level on withdrawal of the electrodes from the cell was always excellent to within 3 mV, even after exposure to the low intracellular $Ca^{2+}$-level for an hour or more. As shown in Fig. 2A, there was no slow phase on withdrawal comparable to that on electrode insertion.

Although the slope of the electrodes is less than Nernstian in this region of $[Ca^{2+}]$, the response is by no means saturated as it was the case with the older electrodes. We calculated an intracellular calcium activity of less than 10 nM from the modified Nernst equation, accounting for the single-ion activity coefficient, of 0.25 for the calcium in the extracellular solution (4). This is considerably lower than values reported before from measurements with electrodes employing different types of ion-exchangers (9, 21).

Appreciable increases in submembrane $Ca^{2+}$-activity from the thus established resting activity can be calculated for voltage-dependent $Ca^{2+}$-influx through the membrane. Published data on calcium-current density mentioned before, together with the diffusion coefficient for free $Ca^{2+}$ of $6 \times 10^2$ $\mu m^2$/s (13) and the expression for diffusion during constant influx into the semi-infinite space (8) yield values of 30 $\mu M$ after 20 ms, (the usual duration of an action potential in these cells). Such values of $[Ca^{2+}]$ should result in large signals in $V_{Ca}$ with either type of electrode, but traces in Fig. 2B, show that this was not the case. Detectable rises in $[Ca^{2+}]_i$ are only obtained with very sustained depolarizations, even then they are quite small. An explanation for this is that the $Ca^{2+}$-selective electrode detects the local calcium-activity just at its very tip, and this will normally be located at a distance of some 10 $\mu m$ or more from the excitable membrane. It seems that intracellular $Ca^{2+}$-sequestration creates steep internal gradients when $Ca^{2+}$-influx through the membrane occurs, quite similar to the gradients directly observed with image-intensifier techniques and $Ca^{2+}$-injection induced aequorin-luminescence (22). The major limitation for the intracellular application of $Ca^{2+}$-selective electrodes to the investigation of excitation-coupled changes in internal calcium thus is not due to insufficient response of the electrodes, but stems from the inhomogeneous $Ca^{2+}$-distribution in the cell under these conditions.

For an investigation of $Ca^{2+}$-dependent $K^+$-activation in spite of these difficulties a more homogeneous $Ca^{2+}$-distribution is obviously needed. We accomplished this by the fast injection of relatively large quantities of $CaCl_2$ solution, so that the injected ions spread across the cell before they are finally sequestered. The basic result of these experiments is shown in Fig. 3A. It is essential for the interpretation that the injection is an instantaneous event (indicated by the arrow) on the time scale as shown, and that the changes in $V_{Ca}$ and $I_M$ reflect the relaxation from this momentary disturbance of the resting state. In these experiments the $Ca^{2+}$-injecting and the $Ca^{2+}$-sensing pipettes penetrated the cells from about opposite sides, so that the injected calcium sensed by the electrode first had to cross the cell. This shows as a delay in onset of the $V_{Ca}$ signal.

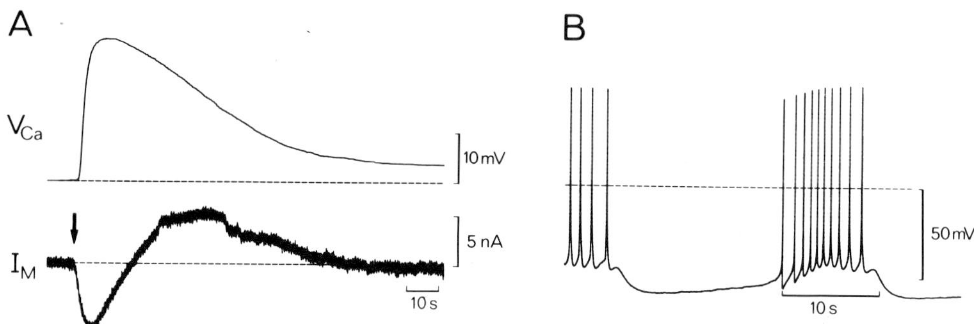


Fig. 3. A) Response of $Ca^{2+}$-selective electrode and of membrane current to an injection of $CaCl_2$ at a constant membrane potential of -50 mV. Time of injection is indicated by arrow. B) Spontaneous discharge pattern of bursting pacemaker neuron. Note biphasic time courses of discharge frequency and spike-afterhyperpolarization during the burst and late minimum of post-burst hyper-polarization

A well defined point to quantify the delay of $V_{Ca}$ from the injection pulse is the point of maximal curvature in its rising phase. By this definition a mean delay of $1.3 \pm 0.7$ s (n = 9) was obtained. Applying the diffusion equation (8) under the assumption of a mean tip separation of both pipettes of 150 um a diffusion coefficient very close to the value of $6 \times 10^2$ um²/s was calculated, which corresponds to free $Ca^{2+}$-diffusion in aequeous media (13). The internal $Ca^{2+}$-binding mechanisms, which slow down effective $Ca^{2+}$-diffusion in flux measurements (6) are probably saturated during this initial phase of a high calcium load and thus do not impede free diffusion. However, the peak internal $Ca^{2+}$-activity sensed 10 s later is only about 5% of the total calcium injected, if homogeneous distribution throughout the cell at this moment is assumed. This means that after this time the large-capacity uptake system with relatively low affinity (3) has sequestered 95% of the injected calcium. While this sequestering continues, $V_{Ca}$ falls with a mean half-time of 24 s. This recovery period showed the largest scatter in the different experiments, with a range from 7 to 45. When injections into the same cell were repeated, there was a tendency for slower recovery from the later injections.

The simultaneously recorded membrane current displays the development of an inward transient which starts without measurable delay from the injection pulse. The roughly similar time course of this inward current with that of the $V_{Ca}$-signal suggests that it is due to direct action of the injected calcium ions, as they spread throughout the cell and reach an increasingly larger part of the internal membrane surface. The peak of the inward transient

always occurred somewhat earlier than that of $V_{Ca}$, at least partly due to
the subsequent activation of an oppositely directed outward current, carried
by the $K^+$-conductance originally observed and described by Meech (19). The
peak of this secondary effect of calcium-injection on membrane current occurs
clearly rather late during the phase of declining $V_{Ca}$. By pulsing the membrane
to different potentials during the two phases of current activation the currents' equilibrium potentials can be calculated and it becomes possible to
separate the two conductances carrying the inward and the outward currents.
It could thus be shown that the delay in activation of the outward current is
not just apparently produced by the transient inward current, but is due to
a real delay in $g_K$-activation. This means that no direct proportionality
exists between $\left[Ca^{2+}\right]_i$ and $g_K$, and that they are probably connected by one
or more intermediate steps.

Control injections of 100 mM KCl, pure water or the ion-exchanger in the injection pipette were without any effect on membrane current, excluding an ar-
tefactual origin of the observed inward-outward current sequence from our injection technique. As a by-product it was confirmed that ion-exchanger does
not react intracellularly and does not produce electrically observable al-
terations of membrane properties. If sufficient time was allowed for the neurons to recover from each injection - and if the injection pipette did not
block in the course of the experiment - the sequence of inward-outward current responses can be elicited by repeated injections for an arbitrary number of times. In one such experiment lasting for several hours calcium was
injected 42 times with qualitatively identical effects, but some slowing of
the time courses at later times.

These findings are able to clarify the origin of the spontaneous activity
pattern of bursting pacemaker neurons, shown in Fig. 3B. One of the most pro-
minent features of this activity is the biphasic time course of the discharge
frequency during each burst. The primary increase in the rate of action potentials is hard to understand on the basis of a $K^+$-conductance which is in
direct proportion to submembrane calcium-levels. It is established from measurements of calcium-increases by optical methods, either using the photopro-
tein aequorin (23) or the metallochromic dye Arsenazo III (10) that $\left[Ca^{2+}\right]_i$
starts to rise with the first action potential of each burst. Using our observation of an inward current, activated in proportion to the submembrane
calcium level, which increases throughout the burst (10, 23), plus the delayed activation of a hyperpolarizing process, bursting discharges are a natural consequence. It also explains the fact that the maximum of the post-
burst hyperpolarization is delayed by some 5 s from the last action potential of the burst, and thus from the absolute peak in internal calcium activity.


Summary

Our experiments have shown that single-barrelled electrodes made from ETH
1001 ion-exchanger are able to faithfully record the extremely low intracellular calcium levels in snail neurons. For the measurement of changes in
$\left[Ca^{2+}\right]_i$ their applicability is limited to slow processes, which maintain a
homogeneous intracellular $Ca^{2+}$-distribution. This is, however, not due to
limited sensitivity of the electrodes, but a result of the local measurement
they perform and the unique role of calcium as an intracellular messenger.
Fast transient changes in neuronal calcium level accompanying electrical activity are better resolved by optical methods. Only the combination of both

methods gives complete information on absolute levels with well resolved transient changes.

It was found that the direct effect of increased calcium levels on the excitable membrane is an opening of channels which can carry inward current and thus contribute to depolarization. The better known calcium-mediated $K^+$-conductance is only subsequently activated, and one or more intermediate reactions are postulated to account for this delay. The local measurement of $[Ca^{2+}]_i$ with an intracellular $Ca^{2+}$-selective microelectrode was essential for the interpretation of the voltage-clamp data. Bursting pacemaker activity occurs as a direct consequence from the observed conductance sequences.


## References


1. Ammann D, Gueggi M, Pretsch E, Simon W (1975) Improved calcium selective electrode based on a neutral carrier. Analyt Lett 8:709-720
2. Ashley CC, Campbell AK (1979) Detection and measurements of free $Ca^{2+}$ in cells. Elsevier/North-Holland Biomedical Press, Amsterdam New York Oxford
3. Baker PF, Schlaepfer WW (1978) Uptake and binding of calcium by axoplasm isolated from giant axons of Loligo and Myxicola. J Physiol (Lond) 276:103-125
4. Bates RG, Staples BR, Robinson RA (1970) Ionic hydration and single ion activities in unassociated chlorides at high ionic strengths. Anal Chem 42:867-871
5. Bjerrum J, Schwarzenbach G, Sillen LG (1957) Stability constants. Part I. Organic Ligands. The Chemical Society, London
6. Blaustein MP, Hodgkin AL (1969) The effect of cyanide on the efflux of calcium from squid axons. J Phyisol (Lond) 200:497-527
7. Brown AM, Brown HM (1973) Light response of a giant Aplysia neuron. J Gen Physiol 62:239-254
8. Carslaw HS, Jaeger JC (1959) Conduction of heat in solids. Clarendon Press, Oxford
9. Christoffersen GRJ, Simonsen L (1977) Ca-sensitive microelectrode: intracellular steady state measurement in nerve cell. Acta Physiol Scand 101:492-494
10. Gorman ALF, Thomas MV (1978) Changes in the intracellular concentration of free calcium ions in a pace-maker neurone, measured with the metallochromic indicator dye arsenazo III. J Physiol (Lond) 275:357-376
11. Heinemann U, Lux HD, Gutnick MJ (1977) Extracellular free calcium and potassium during activity in the cerebral cortex of the cat. Exp Brain Res 27:237-243
12. Heyer CB, Lux HD (1976) Properties of a facilitating calcium current in pacemaker neurones of the snail Helix pomatia. J Physiol (Lond) 262: 319-348
13. Hodgkin AL, Keynes RD (1957) Movements of labelled calcium in squid giant axons. J Physiol (Lond) 138:253-281
14. International Union of Pure and Applied Chemistry (1976) Recommendations for nomenclature of ion-selective electrodes. Pure and Appl Chem 48: 127-132
15. Kostyuk PG, Krishtal OA (1977) Separation of sodium and calcium currents in the somatic membrane of mollusc neurones. J Physiol (Lond) 270: 545-568
16. Kretsinger RH (1979) The informational role of calcium in the cytosol. Adv Cyclic Nucleotide Res 11:1-26

17. Llinas R, Blinks JR, Nicholson C (1972) Ca-transient in presynaptic
    terminal of squid giant synapse: Detection with Aequorin. Science
    176:1127-1129
18. Lux HD, Neher E (1973) The equilibration time course of $K^+$ in cat cor-
    tex. Exp Brain Res 17:190-205
19. Meech RW (1972) Intracellular calcium injection causes increased potas-
    sium conductance in Aplysia nerve cells. Comp Biochem Physiol 42:
    493-499
20. Meech RW (1979) Membrane potential oscillations in molluscan "burster"
    neurones. J Exp Biol 81:93-112
21. Owen JD, Brown HM, Pemberton JP (1977) Neurophysiological applications
    of a Ca-selective microelectrode. Analyt Chim Acta 90:241-244
22. Rose B, Loewenstein WR (1975) Calcium ion distribution in cytoplasm vi-
    sualized by Aequorin: Diffusion in cytosol restricted by energized
    sequestering. Science 190:1204-1206
23. Stinnakre J, Tauc L (1973) Calcium influx in active Aplysia neurones de-
    tected by injected Aequorin. Nature New Biol 242:113-115
24. Tillotson D (1979) Inactivation of Ca conductance dependent on entry of
    Ca ions in molluscan neurons. Proc Natl Acad Sci 76:1497-1500

Max-Planck-Institut für Psychiatrie, Abt. Neurophysiologie, Kraepelinstr. 2,
8000 München 40 / FRG

Discussion

Nicholson:  Did the increase in cell volume due to calcium injection introduce any artefact into the voltage clamp?
Hofmeier:  Usually we injected an amount equal to only 1% of the cell volume.
Control injections with distilled water or oil up to as much as 15% of the
cell volume, did not produce any significant artefacts.

# Effects of Lithium Application Upon Extracellular Potassium Structures of the Peripheral and Central Nervous System of Rats

G. TEN BRUGGENCATE, A. ULLRICH, M. GALVAN, H. FÖRSTL, P. BAIERL

The mechanism of the therapeutic action of lithium (Li) is still unclear. Although it has been proposed that Li acts on specific transmitter systems (3) we felt that the primary effect is a more general neurobiological action. In particular, one factor involved may be an influence upon other ions which are important for neuronal functions. It has already been shown that Li can inhibit the electrogenic Na/K pump in the isolated vagus nerve (7,8). However, since in this and subsequent studies high concentrations of Li were used, the clinical relevance of this finding has been questioned (9). We have reinvestigated this problem with ion selective microelectrodes (ISME); particular attention was paid to the effects of acutely applied low Li concentrations. This has been compared with effects observed after chronic oral administration.

Unmyelinated fiber preparations were used since changes in ion balance ought to show up most clearly in small neuronal elements having a large surface/volume ratio. We compared the isolated superior cervical ganglion, the isolated vagus nerve, and the cerebellar cortex in vivo. Poststimulus $K^+$-undershoots were taken as an indicator of Na/K pump activity (5).

## Methods

Liquid ion-exchanger microelectrodes were used to monitor the extracellular activity of $K^+$ (Corning 477317) and $Li^+$ (see 10), respectively.

### In Vitro Preparations:

Cervical vagus nerves or superior cervical ganglia (4) were isolated from urethane-anaesthetized rats and maintained in flowing Krebs' solution ($25^oC$, K = 6, Ca = 2.5 mmol/l). The stimulus strength was adjusted for vagus nerves to evoke maximal C-fiber volleys; LiCl was added to the Krebs' solution.

### In Vivo Preparations:

Recordings were made within a parallel fiber beam, activated by cerebellar cortical surface stimulation (6) in urethane anaesthetized rats (recording depth 100 /um). The cerebellum was kept moist by continuous superfusion with a modified Krebs' solution at $37^oC$. In one series of experiements LiCl was acutely applied via the superfusate (exchange for NaCl). In another series, male albino rats received 30 and 100 mmol LiCl per kg dry food for periods between 3 weeks and 3 months. The two groups had plasma Li concentrations of about 0.5 and 1.0 mmol/l, respectively at the time at which the actual measurements were performed.


## Results

In general, the result of acute or chronic Li application was that the resting $K^+$-level became elevated and the poststimulus $K^+$-undershoots reduced (see also 11).

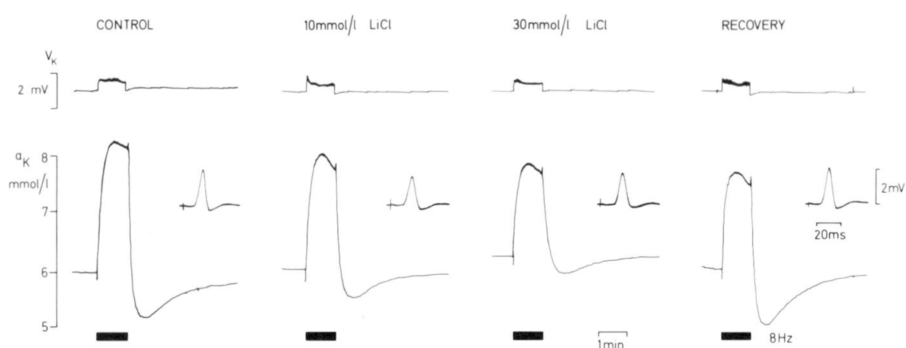


Fig. 1.   Effects of Li upon extracellular $K^+$ in an isolated superior cervical ganglion
Stimulation at 8 Hz, as indicated by black bars, raised the $K^+$-level from 6 up to more than 8 mmol/l (control). the insets show the compound potentials recorded with a suction electrode from the postganglionic nerve; the top line is the dc field potential recorded from the reference barrel of the ISME. Addition of LiCl to the Krebs' solution resulted in slightly reduced $K^+$-peak values during stimulation, reduced $K^+$-undershoots after the end of stimulation, and an elevation of the prestimulus $K^+$-level. Upon washout (=recovery), the prestimulus $K^+$-level returned to normal and the $K^+$-undershoot was enhanced


## Acute Li-Administration:

Figure 1 illustrates representative data obtained in isolated ganglia. The resting $K^+$-level was 6 mmol/l corresponding to the $K^+$-concentration of the Krebs solution. Upon stimulation of the preganglionic nerve, $K^+$ rose up to a peak of about 8 mmol/l and undershot the baseline value after the end of the stimulation period. During 10 mmol/l LiCl application, the poststimulus undershoot was reduced by about 40%. At 30 mmol/l this effect was even larger and accompanied by a rise in resting $K^+$-level. The interpretation of the reduced undershoot is complicated by the fact that the stimulus induced $K^+$-peak is somewhat reduced (particularly at the high $[Li^+]$ ), probably because of reduced synaptic transmission as indicated by the diminished postsynaptic population potential (inset). However, upon washout, the undershoot was enhanced whereas at the same time the $K^+$-peak was certainly not increased. Similar results were obtained in the isolated vagus nerve preparation, although quantitatively an undershoot reduction comparable to the one seen with 10 mmol/l $Li^+$ in the ganglion occurred only at about 30 mmol/l.

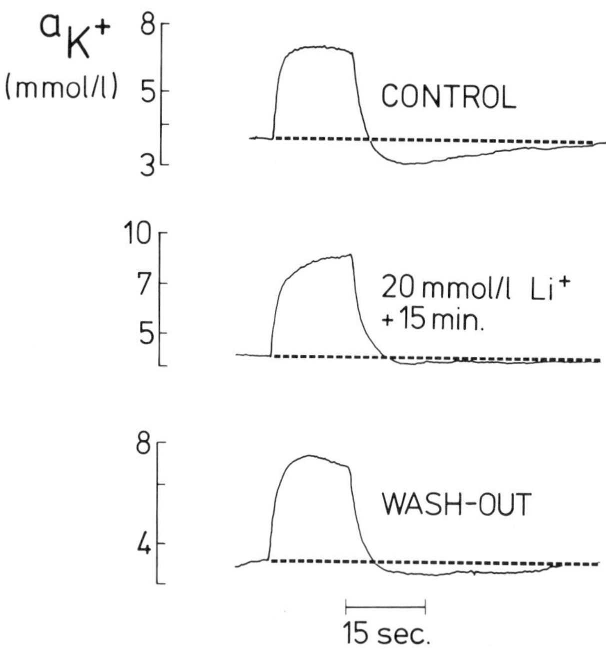



Fig. 2. Effects of $Li^+$ upon extracellular $K^+$ in cerebellar cortex
Stimulus-induced $K^+$-transients (12 Hz) similar to Figure 1. Superfusion with $Li^+$ (20 mmol/l) resulted in an elevation of the resting $K^+$-level by about 1 mmol/l, and a clear reduction of the $K^+$-undershoot. The effects were reversible upon washout

Figure 2 illustrates coresponding findings obtained in the cerebellar cortex. In the experiment illustrated, the actual $Li^+$-concentration within the cerebellar cortex was not measured. However, the mean value of $Li^+$ obtained in other experiments during superfusion with 20 mmol/l was 7.4 mmol/l (100 /um depth, 15 min. after start of $Li^+$-superfusion).
As seen in Fig. 2, the resting $K^+$-concentration rose from about 3.5 to 4.5 mmol/l, and the poststimulus $K^+$-undershoot became clearly reduced during $Li^+$-superfusion.

Table 1 summarizes the $K^+$-parameters during acute and chronic $Li^+$-application. In the upper part of Table 1, 4 groups (A-D) are classified according to various Li-concentrations reached during Li-superfusion periods (aLiCb = extracellular $Li^+$-concentration within the cerebellar cortex as measured with LiISMEs). The elevation of the resting $K^+$-levels (aKCb), already shown in Fig. 2, appears to be dose-dependent. Despite an increase in the stimulus-induced $K^+$ peak values ($\Delta Kp$), the $K^+$-undershoots ($\Delta Ku$) were reduced during $Li^+$ superfusion.

Table 1. Changes in extracellular potassium related to lithium application

| | aLiCb<br>mmol/l | aKCb | $\Delta$Kp<br>percentage of controls | $\Delta$Ku |
|---|---|---|---|---|
| A | 1.6+0.2 | 106+1<br>** | 106+1 | 97+2 |
| B | 5.3+0.5 | 112+1<br>*** | 116+1<br>*** | 87+2<br>*** |
| C | 7.4+0.3 | 121+1<br>*** | 146+2<br>*** | 60+2<br>*** |
| D | 20.0+2.0 | 181+10<br>** | + | + |
| E | 0.7+0.3 | 106+2<br>* | 92+4 | 80+11 |
| F | 1.3+0.3 | 119+2<br>*** | 104+7 | 74+6<br>* |

A-F: groups of rats with different extracellular $Li^+$-concentrations within the cerebellar cortex. The various $Li^+$-levels (aLiCb) were obtained by acute superfusion with various concentrations (A-D), or by chronic oral administration of different Li-diets (E,F). The latter two groups had plasma $Li^+$-values of 0.5+0.1 and 1.0+0.2 mmol/l, respectively. Potassium parameters were normalized with respect to control values (100%). In the acute experiments, there was a dose-dependent increase in the resting $K^+$-level and the stimulus-induced peak, whereas the $K^+$-undershoot was reduced. Qualitatively similar changes were observed in chronic rats; no change was seen in the stimulus-induced peak. All values: means $\pm$ SEM. n: between 5 and 20 in the various measurements.

     * = p 0.05
    ** = p 0.01
   *** = p 0.001
    + = at this extracellular [$Li^+$] inactivation occurred, precluding measurements of stimulus-induced $K^+$ changes.


Chronic Oral Administration

In rats fed with a Li-diet and having plasma $Li^+$-concentrations of 0.5 and 1.0 mmol/l, respectively, the cerebellar levels (measured with ISMEs) were about 0.7 and 1.3 mmol/l, respectively (E and F). Also, in these animals resting $K^+$ was elevated (groups E and F) and $K^+$-undershoots reduced (F).


Cardiac Glycosides

Both an elevation of the extracellular resting potassium ion concentration and a decrease in poststimulus $K^+$-undershoots are compatible with a reduction of the Na/K pump activity. In fact, K-strophanthidin also increased the $K^+$-level and decreased $K^+$-undershoots. In addition, these effects of cardiac glycosides were more pronounced in animals during either acute $Li^+$-superfusion or chronic Li-treatment.


## Discussion


In all three preparations studied the actions of $Li^+$ were consistent with an inhibition of the Na/K pump (see also 7). Thus, the resting $K^+$-level was elevated and the amplitude of poststimulus $K^+$-undershoots was reduced. These effects were qualitatively similar to those produced by cardiac glycosides (see also 4), although the site of action of $Li^+$ within the pump process cannot be determined from our study. In the in vivo experiments, an acute application of 6 mmol/l $Li^+$ had actions quantitatively similar to those observed after chronic treatment, which produced a plasma $Li^+$ concentration of only 1 mmol/l. The parameter identical in both cases could be the intracellular $Li^+$ concentration, which we have not yet measured.

An impairment of the transport capacity of the Na/K pump is expected to become most evident in the smallest neuronal elements, since normal ion concentrations in structures with a small intracellular volume are particularly dependent upon pump activity. The specificity of clinical $Li^+$-effects with respect to endogenous depression could be due to such a preferential expression of $Li^+$ actions among the smallest neuronal components. In small diameter axons and at synaptic terminals, the inhibition of the sodium pump may lead to an elevation of the intracellular sodium (and eventually also calcium) concentrations. The most important consequence would be an enhanced transmitter release (2), which has been postulated to be important for long-term synaptic facilitation (1). Higher concentrations of $Li^+$ probably lead to a reduced sodium ion gradient across the membrane and an inactivation of sodium channels. This could explain the toxic actions of $Li^+$, which may be analoguous to the cardiotoxic effects of glycosides on the Na/K pump.


## References


1. Atwood HL, Swenarchuk LE, Gruenwald CR (1975) Long-term synaptic facilitation during sodium accumulation in nerve terminals. Brain Res 100: 198
2. Birks RI, Cotten MW (1968) The influence of internal sodium on the behaviour of motor nerve endings. Proc Roy Soc London B 170: 401
3. Bunney WE, Murphy DL (eds) (1976) The neurobiology of lithium. Neurosciences Res Prog Bull 14: 111
4. Galvan M, ten Bruggencate G, Senekowitsch R (1979) The effects of neuronal stimulation and ouabain upon extracellular $K^+$ and $Ca^{2+}$ levels in rat isolated sympathetic ganglia. Brain Res 160: 544
5. Heinemann U, Lux HD (1975) Undershoots following stimulus induced rises of extracellular potassium concentration in cerebral cortex of cat. Brain Res 93: 63

6. Nicholson C, ten Bruggencate G, Steinberg R, Stöckle H (1977) Calcium modulation in brain extracellular microenvironment demonstrated with ion-selective micropipette. Proc Nat Acad Sci (Wash) 74: 1287
7. Ploeger E (1974) The effects of lithium on excitable cell membranes. On the mechanism of inhibition of the sodium pump of non-myelinated nerve fibres of the rat. Europ J Pharmacol 25: 316
8. Ritchie JM, Straub RW (1957) The hyperpolarization which follows activity in mammalian non-myelinated fibres. J Physiol 136: 80
9. Schou M (1976) Pharmacology and toxicology of lithium. Ann Rev Pharmacol Toxicol 16: 231
10. Thomas RC, Simon W, Oehme M (1975) Lithium accumulation by snail neurones measured with a new $Li^+$-sensitive microelectrode. Nature 258: 754
11. Ullrich A, Baierl P, ten Bruggencate G (1980) Extracellular potassium in rat cerebellar cortex during acute and chronic lithium application. Brain Res 192: 287

Department of Physiology, University of Munich, Pettenkoferstr. 12, 8000 München 2, FRG

# Micro-Electrode Measurement of Skin pH in Post-Operative Intensive Care Patients

D.K. HARRISON, W.F. WALKER

## Summary

Experiments have been carried out on normal volunteers to investigate the effects of ischaemia, hypoxia and local temperature changes on skin pH as measured with micro-electrodes. In view of the knowledge gained from these experiments, the study was extended to measurements in patients with peripheral vascular disease and to post-operative intensive care patients. The results show that skin pH can reflect the acid-base status, but that the over-riding effect is that of peripheral ischaemia caused either by shock or arterial occlusion.

## Introduction

The possibility of using the measurement of tissue pH as an early warning system for shock in critically ill patients was first suggested more than a decade ago (6). Since that time several different types of electrode have been designed to enable the surface pH of skeletal muscle to be monitored (5, 8), and results have shown (1) that muscle surface pH is a good indicator of peripheral perfusion. The main disadvantage of this technique is that it is highly invasive; even the smallest of the electrodes used is of the order of 0.4 mm in diameter. In order to overcome this drawback we decided, in the light of experience gained from tissue $PO_2$ measurements (11), to design an electrode which would enable us to monitor local extracellular pH changes in skin.

## Skin pH Measurement

A relatively large electrode has been in use for some time (12) for monitoring the subcutaneous tissue pH of foetuses, newborn infants and, more recently, adults (9). However, we have designed and constructed (2) a skin pH micro-electrode which has a tip diameter of the order of only 50 um, but which is strong enough to withstand the stresses involved in inserting it into skin. Indeed, our internally insulated glass micro-electrodes can be re-used many times.

The characteristics of the micro-electrode and the technique used for measuring skin pH have been fully described elsewhere (3, 4). Briefly, the technique involves the insertion of up to five pH micro-electrodes in a lower limb of the subject (or patient), together with a single reference micro-electrode (Fig. 1). The electrodes are calibrated in phosphate buffers at $33^{\circ}$C both before and after each experiment. After the electrodes have been sterilized, they are inserted into the skin to a depth of about 5 mm at an angle of approximately $30^{\circ}$. A stabilisation period of about 10 min is allowed before starting to record pH values. Any abnormal skin temperature which may be observed in the region of the electrodes are compensated for in the post-experimental calibration.

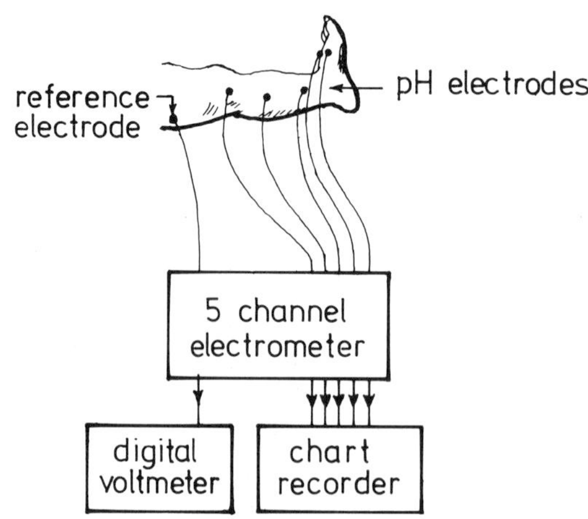



Fig. 1. The 5-channel skin pH recording system


## Skin pH Changes in Normal Subjects

Three series of experiments were carried out on normal volunteers to investigate the effects on skin pH of respiratory hypoxia, ischaemia, and local temperature changes (3). All of these produced significant changes in skin pH. For example, Fig. 2 illustrates the effect of breathing a 10% $O_2$ in $N_2$ mixture for a 10 min period. Hyperventilation during the period of hypoxia caused a mean increase of pH 0.04 $\pm$ 0.02 (SD) in skin pH in 10 volunteers. Subsequently, skin pH fell to a value 0.02 $\pm$ 0.02 pH units below normal following the hypoxic period, suggesting the presence of excess lactate. Skin pH results were comparable with the blood gas and pH values of arterialized samples taken throughout each experiment.

A 20 min period of tourniquet ischaemia in 20 volunteers induced a fall in skin pH of 0.13 $\pm$ 0.05 pH units. This was equivalent to a 32% increase in $H^+$ ion activity.

The largest changes in skin pH were recorded whilst investigating the effect of cooling the area of skin in which two electrodes were inserted. A $10^\circ$C fall in temperature produced a mean increase in skin pH of 0.23 $\pm$ 0.07 pH units in the 10 volunteers studied.

The mean value for the pH in human skin, determined from 160 electrode recordings during 44 experiments on 40 different normal volunteers aged between 19 and 30 years, was found to be pH 7.54 $\pm$ 0.09. No difference in skin pH was observed between males and females, or in different regions of the limb. This value may appear to be surprisingly high. However, it should be remembered that all other pH values, including that of blood, are reported at a standard temperature of $37^\circ$C, whereas skin pH is measured in the dermis at a temperature estimated thermographically to be $33^\circ$C: this is taken to be

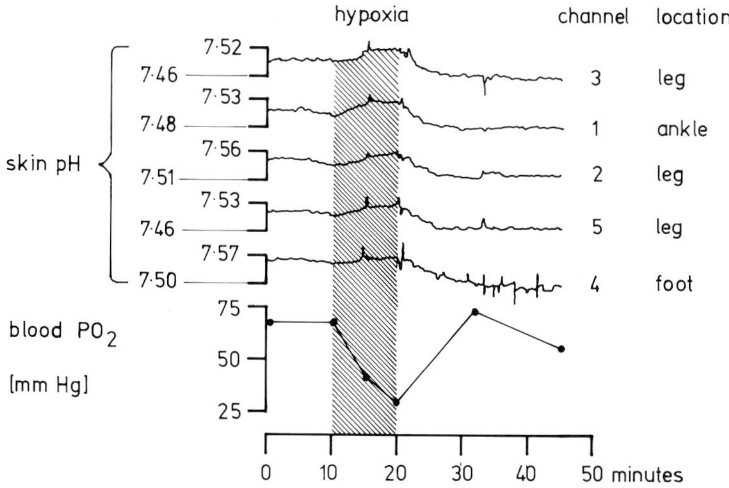


Fig. 2.  Normal subject E.H. - Skin pH and arterialized blood $pO_2$ values during hypoxia

equivalent to a skin surface temperature of $30^{\circ}C$. Thus, it would be valid, for comparison purposes, to correct the skin pH value by the measured factor (see above) of 0.023 pH units per $^{\circ}C$ rise in temperature. The mean normal skin pH value would therefore become pH 7.45 $\pm$ 0.09 at the standard temperature of $37^{\circ}C$.

## Skin pH Measurements in Patients

Having established the normal value of skin pH, and the magnitude of its response to changes in blood flow, oxygen supply and temperature, it was appropriate to extend the use of the skin pH micro-electrode to certain clinical situations. The ultimate aim in designing the micro-electrode, as stated earlier, was to monitor patients in shock. However, it was considered that the measurement of skin pH in the ischaemic lower limbs of patients with occlusive arterial disease would provide an ideal opportunity to test the system clinically whilst continuing to use it in the laboratory environment. Furthermore, the chronic low blood flow condition which exists in these limbs serves as a model for the acute state which occurs in the shocked patient.

## Patients with Peripheral Vascular Disease

This study was carried out in 11 patients (10 men and 1 woman) with various degrees of lower limb ischaemia. Their ages ranged from 50 to 76 years. In one case, a second measurement was made in the same patient. In each case a thermographic record was available which gave an accurate guide as to the severity of the ischaemia. Some patients had gross necrotic and gangrenous areas, usually in the toes and feet, and were being prepared for amputation.

Other patients showed signs of ulceration or oedema but had no gangrene. In all cases the severe peripheral vascular problems had caused ischaemic changes in the skin.

Almost all of the patients studied displayed considerable temperature gradients, sometimes up to $6^{\circ}C$, along the limb in which the measurements were being made. It was necessary therefore, to standardize all skin pH measurements at a skin surface temperature of $30^{\circ}C$ as in the normal skin pH recordings described in the previous section. Thus, the measured temperature coefficient of 0.023 pH units per $^{\circ}C$ was used to correct all skin pH values recorded at skin surface temperatures other than $30^{\circ}C$. For example, a skin pH value of pH 7.48 measured in the foot of patient W. McL. at $26^{\circ}C$ was corrected to pH 7.39 at $30^{\circ}C$.

Five micro-electrodes were inserted in the affected limb: one in the thigh, two in the calf and two in the foot. The reference micro-electrode was inserted in the mid-calf. Temperature recordings were made as close as possible to the site of insertion of each pH electrode. The whole procedure, from the patient arriving at the laboratory to leaving at the end of the test, took approximately 30 minutes.

The mean value of the skin pH measured in the thighs of these 11 patients was pH 7.54 $\pm$ 0.08, which is exactly the same mean value as that found in normal subjects in all parts of the leg. The mean skin pH value in the feet was pH 7.30 $\pm$ 0.13, which lies well below the range of normal values, and the mean skin pH value in calf was pH 7.49 $\pm$ 0.09. The result of thigh, calf and foot skin pH recordings for each patient were compared using Student's t test for paired values, and the differences between thigh-calf and calf-foot skin pH values were found to be significant at the 0.05 and 0.001 levels respectively.

If one examines these results in terms of $H^{+}$ ion activity, it is interesting that in one case, patient S.S., there was a 330% increase between the thigh and foot. However, the mean skin pH values given above represent an increased $H^{+}$ ion activity of 74% above normal in the feet of these patients.

## Measurements in Patients in the Intensive Care Unit (ICU)

Measurements were taken in 6 patients: 3 males and 3 females aged between 41 and 83 years. All of them underwent some form of cardiovascular surgery after which they were immediately transferred to the ICU. At this state, each patient had an indwelling arterial line, and was being artificially ventilated. Whenever possible, skin pH measurements were made pre-operatively in order to obtain a normal value, but this could only be done in two cases. In three of the other cases, normal skin pH values were obtained about 2 weeks after the operation, by which time the patients had recovered. In the sixth case it was not possible to obtain a normal skin pH value.

An average of four micro-electrodes were inserted in a lower limb of the patient for each skin pH measurement. They were located in the thigh, calf and foot; the reference electrode was inserted in the calf. Arterial blood gas and pH levels were measured at the time of each skin pH recording whilst the patient was under intensive care. A total of fourteen recordings were made in the six patients, and in all cases skin pH values were corrected for any temperature differences along the limb.

The results of the skin pH measurements in these patients, together with arterial blood gas and pH values are given in Table 1.

Table 1.  Skin pH and arterial blood gas and pH values

| Patient | Time of measurement | Skin pH | | | | Arterial Blood | | |
|---|---|---|---|---|---|---|---|---|
| | | Foot | Calf | Thigh | Mean | pH | $pO_2$* | $pCO_2$* |
| H.U. | 1 Day Pre-Op. | 7.30 | 7.34 | – | 7.33 | – | – | – |
| | 1 Hr. Post Op. | – | 7.44 | – | 7.44 | 7.34 | 124 | – |
| | 1 Day Post-Op. | – | 7.53 | – | 7.53 | 7.45 | 145 | 23.9 |
| | 18 Days Post-Op. | 7.40 | 7.35 | – | 7.36 | – | – | – |
| G.S. | 4 Days Pre-Op. | 7.38 | 7.38 | – | 7.38 | – | – | – |
| | 1 Hr. Post-Op. | 7.30 | 7.38 | 7.41 | 7.37 | 7.44 | 158 | 25.6 |
| D.R. | 1 Hr. Post-Op. | 7.02 | 7.14 | 7.20 | 7.13 | 7.40 | 146 | 30.4 |
| | 14 Days Post-Op. | 7.32 | 7.41 | 7.40 | 7.39 | – | – | – |
| B.B. | 1 Hr. Post-Op. | 7.33 | 7.41 | 7.36 | 7.37 | 7.40 | 108 | 28.0 |
| | 1 Day Post-Op. | 7.53 | 7.62 | 7.56 | 7.59 | 7.41 | 82 | 46.2 |
| | 14 Days Post-Op. | 7.61 | 7.66 | 7.57 | 7.63 | – | – | – |
| H.H. | 1 Day Pre-Op. | 7.53 | 7.58 | – | 7.56 | – | – | – |
| | 1 Hr. Post-Op. | – | 7.58 | 7.55 | 7.57 | 7.40 | 153 | 36.0 |
| L.H. | 1 Hr. Post-Op. | 7.37 | 7.50 | 7.54 | 7.49 | 7.37[+] | 132 | 30.9 |

* mm Hg
[+] Mixed venous blood

Patient H.U. underwent a mitral valvotomy operation. It is important to note in examining the skin pH results that this patient suffered from McArdle's syndrome (10) which is a deficiency of the enzyme myophosphorylase. This meant that the patient produced no lactate in the muscle during exercise-induced ischaemia, but instead formed an excess of glycogen. The effect of this on skin pH seems to have been to produce a value some pH 0.2 below normal. Note that for the measurements made 1 day post-operatively the patient was in a state of respiratory alkalosis, and this was reflected in what was for this patient a high skin pH value. Patient H.H. underwent a similar operation, and, again skin pH values reflected the arterial blood value postoperatively.

Patient G.S. had an operation to replace a cardiac pacemaker. There was some thermographic evidence that the patient had an early peripheral vascular problem. This is reflected in the relatively low skin pH values, and the gradient recorded during the post-operative measurement. No significant change in skin pH was observed in response to the high arterial $pO_2$ and low $pCO_2$ values. The arterial pH, although high, lay within the normal range.

Patients D.R. and L.H. both exhibited signs of peripheral vascular disease, as illustrated by the skin pH gradients along their legs. An operation was performed on D.R. to repair an aortic aneurysm. During the operation it was necessary to occlude the aorta for almost 3 hours, producing a build-up of lactic acid. This, in turn, gave rise to the extremely low post-operative skin pH recordings. Patient L.H., who had undergone a mitral valvotomy operation, was in a state of respiratory acidosis at the time of measurement. This was reflected, to a small degree, in the slightly depressed mean skin pH value, but the situation was confused by the patient's peripheral vascular problem.

Patient B.B., an 83 year old male, also underwent an operation to repair an aortic aneurysm. As a result of considerable blood loss, 9 litres of blood were infused into the patient before the first post-operative skin pH measurement. A second measurement was made on the following day in the ICU by which time further loss of blood had been stopped, and the patient was no longer being artificially ventilated. The skin pH value recorded during the first measurement was considerably lower than would be expected in the light of the arterial pH and $pCO_2$ values. This low value was almost certainly due to peripheral ischaemia caused by haemorrhagic shock.


Discussion

The results of the measurements in patients with peripheral vascular disease have been confirmed by further tests using the same technique (7). The presence of skin pH gradients in the legs of these patients is explained by the low levels of skin oxygen tension present (11). As a result, excess lactate accumulates and causes an increase in $H^+$ ion activity in the distal parts of the diseased limb.

The clinical value of these findings is that skin pH can indicate the severity of ischaemia, and it is possible that further work might reveal a particular skin pH value below which the survival of tissue is unlikely. Thus, it may prove to be a useful adjunct to other techniques used to determine amputation levels. A further point worth noting is that the recorded increase in $H^+$ ion activity probably results in a lowering of the body's buffer mechanisms, which could make the patient more susceptible to shock.

Interpretation of the results of skin pH measurements in the six intensive care patients is more complex, since it is probable that more than one physiological disturbance was occurring simultaneously. However, our results (and those of other workers (9)) show that skin pH is primarily sensitive to ischaemia due either to shock or arterial occlusion, and only secondarily sensitive to respiratory changes. In other words, skin pH will be indicative of the respiratory state of a patient only if no ischaemia exists at the site of measurement.

It is difficult to obtain a representative sample of the type of patient being discussed here, so the comparison of pH measurements within individual patients is the only practical way of assessing results. For this reason it is preferable to obtain a pre-operative skin pH measurement in order to interpret immediately those recordings made in the ICU. None the less, our results indicate that skin pH values in shocked patients may lie considerably below the normal range.

It is recognized that our method of skin pH measurement has limitations. For example, for reasons relating to the necessary care of such critically ill patients, it is not possible to leave the micro-electrodes implanted for more than about an hour. This, and other factors weigh against the advantages which may be gained from the technique's routine clinical application. However, the results of our investigations in patients with peripheral vascular disease and in post-operative intensive care patients indicate two areas where skin pH monitoring could be of considerable clinical importance.

## References


1.  Filler RM, Das JB, Espinosa HM (1972) Clinical experience with continuous muscle pH monitoring as an index of tissue perfusion and oxygenation and acid-base status. Surgery 72:23-33
2.  Harrison DK, Walker WF (1977) A new design of glass micro-electrode for extracellular pH measurement. J Physiol 269:23-25P
3.  Harrison DK, Walker WF (1979) Micro-electrode measurement of skin pH in humans during ischaemia, hypoxia and local hypothermia. J Physiol 291: 339-350
4.  Harrison DK, Walker WF (1980) Tissue pH electrodes for clinical applications. J Med Eng Technol 4:3-7
5.  Kung TLW, LeBlanc Jr OH, Moss G (1976) Percutaneous microsensing of muscle pH during shock and resuscitation. J Surg Res 21:285-289
6.  Lemieux MD, Smith RN, Couch NP (1969) Electrometric surface pH of skeletal muscle in hypovolemia. Am J Surg 117:627-631
7.  Meehan SE, Walker WF (1979) Measurements of tissue pH in skin by glass micro-electrodes. Lancet II:70-71
8.  O'Donnell Jr TF (1975) Measurement of percutaneous muscle surface pH. Lancet II:533
9.  Rithalia SVS, Herbert P, Tinker J (1979) Continuous monitoring of tissue pH. Brit Med J I:1460
10. Salter RH, Adamson DG, Pearce GW (1967) McArdle's syndrome (myophosphorylase deficiency). Q J Med 36:565-578
11. Spence VA, Walker WF (1976) Measurement of oxygen tension in human skin. Med Biol Eng Comput 14:159-165
12. Stamm O, Latscha U, Janacek P, Campana A (1973) Kontinuierliche subkutane pH-Messung am kindlichen Kopf sub partu und post partum. Gynaek Rundsch 14:28-35


## Acknowledgements


We are grateful to the Equipment Research Committee of the Scottish Home and Health Department for funding this project.


The University, Dundee DD1 4 HN/Scotland


## Discussion

Lübbers:  Similarly as the transcutaneous $PO_2$ or $Pco_2$ measurements, the intracutaneous pH measurements depend very much on local circulation. It is known that even a small skin damage may change local circulation considerably. Thus, variation in local circulation may influence your results. Therefore, to obtain meaningful results it seems necessary to find a way to record local circulation simultaneously.

# Intracellular pH, Na$^+$ and Ca$^{2+}$ Activity Measurements in Mammalian Heart Muscle

D. ELLIS, J.W. DEITMER*, D.M. BEERS**

## Introduction

Ion-selective microelectrodes are now available for many ionic species so that studies are now feasible on the effects of changes in the activity of one type of ion on the activity of other ions in the cell. From initial studies on the control of intracellular Na$^+$ activity (aNa$_i$) in cardiac tissues we became interested in transmembrane Na$^+$-Ca$^{2+}$ exchange. From our experiments we would have predicted that under some conditions a reduction of extracellular Na$^+$, ([Na]$_o$), would produce large charges in extracellular Ca$^{2+}$ activity (aCa$_i$). These changes in aCa$_i$ would however also probably change intracellular pH (pH$_i$) as it has been shown that pressure injection of Ca$^{2+}$ into snail neurones results in intracellular acidification (10). Therefore under conditions that produce large changes in internal Ca$^{2+}$ the pH$_i$ would also be expected to alter. Changes in aCa$_i$ could be demonstrated by the use of Ca$^{2+}$-selective microelectrodes and by measuring the tension that the muscle developed.

This type of interaction between ions may be very important in heart muscle. The strength of contraction is of course Ca$^{2+}$-dependent but changes in pH also affect contractile activity. The aNa$_i$ also influences contractile strength, probably largely due to Na$_i^+$-Ca$_o^{2+}$ exchange at the cell membrane. High doses of cardioactive steroids inhibit the Na$^+$-K$^+$ pump and cause a rise of aNa$_i$. Concomitant with this rise of aNa$_i$ is the development of a contracture (19). We have been particularly interested in this latter mechanism due to the importance of cardioactive steroids in potentiating contractile strength in heart muscle. In order to interpret the results of some types of experiments it is therefore necessary to observe not only the primary effect of a given intervention but also to look for secondary changes, as these might also have very important influences on the properties of the tissue.

## Methods

The methods and experimental set up have been previously described (2-7). Recessed-tip Na- and pH-sensitive microelectrodes have been used (14, 15, 17). The microelectrodes are well suited for continuously monitoring aNa$_i$ and pH$_i$ in mammalian cardiac tissues over long time periods. Providing the electrode tips are 1/um or less in diameter they penetrate cells fairly readily and ensure long stable measurements. The recessed volume can be made small enough for the response times of the electrodes to be well within the time courses of most changes in intracellular ion levels. They cannot, however, detect the rapid changes that occur during single action potentials. Most of our work has been carried out on quiescent preparations as we have been interested primarily in membrane transport processes. It is possible to estimate the effects of electrical (and contractile) activity of the preparations by stimulating the preparations for periods of ten or fifteen minutes, then stopping the stimulation and observing the level reached of a particular ion, and measuring the rate of recovery of that ion activity. Experiments of this type have shown that $aNa_i$ increases by only 1-3 mM following this type of treatment (4) and $aNa_i$ recovers with a time course indicative of the action of the $Na^+-K^+$ pump (3). In general we have avoided the use of contractile preparations for several reasons: (a) The ion flow during the action potential can change intracellular ion levels and produce unneccessary complications when one is trying to study trans-membrane ion exchange under the simplest conditions. (b) Sustained micro-electrode impalement is more difficult in vigorously contracting tissues. This is not a great problem in Purkinje fibre measurements where the force of contraction is normally weak but is a problem in other cardiac tissues. The recessed-tip design of ion-selective microelectrode with its double-glass walled structure renders the tip relatively inflexible so that main-tenance of stable penetrations is more difficult. (c) The high resistance and capacitance of the ion-selective glass membranes result in a consider-able attenuation and slowing of the recorded action potentials. The ability to record an intracellular ion level adequately is dependent on subtraction of the membrane potential, recorded by a conventional microelectrode, from the signal recorded by the ion-selective microelectrode. At physiological rates of stimulation the ion-selective microelectrode is often too slow to be able to follow the rapid changes in membrane potential accurately. This would lead to inaccurate estimates of intracellular ion levels.

The $Ca^{2+}$-selective microelectrodes that we used employed the neutral ligand ETH-1001 (13). Such electrodes have much faster response times than the recessed-tip $Na^+$- and pH-selective glass microelectrodes. The micropipettes were initially silanized by dipping their tips in tri-n-butylchlorosilane (2.5% in 1-chloronaphthalene) for approximately 15 sec and then drying in an oven ($80^\circ$C) for 2 hours. They were filled through the tip by dipping the tip in the solution of the ligand. Gentle suction (from a 2.5 ml syringe) was applied to the back end. Filling normally took about 30 min for a column $100-150\,\mu$m in length in a pipette that would have a resistance equivalent to about 5 M$\Omega$ if filled with 3 M KCl. The electrodes were cali-brated in Ca-EGTA solutions using a binding constant of 3.2 x $10^6$/M at pH 7.00, $22^\circ$C. The size of the response to $Ca^{2+}$ changes was small over the normal range of intracellular $Ca^{++}$ levels but we were particularly interest-ed in the very large $Ca^{2+}$ changes that occur when changes of $aNa_i$ and $pH_i$ are produced.

When looking at changes in the activity of an ion in a cell induced by changes in the activity of other ions, care must be taken to ensure that artifacts are not introduced simply because the ion-selective microelec-trodes respond to more than one ionic species. For all three types of ion-selective microelectrode used the apparent changes in intracellular ion levels measured could not be accounted for by interference from changes in other ion levels.


## Results and Discussion

Cardioactive steroids like ouabain and strophanthidin at concentrations of about $10^{-7}$M cause an increase of $aNa_i$ in sheep heart Purkinje fibres as a result of inhibition of the $Na^+-K^+$ pump at the cell membrane (4, 5). Concentrations of strophanthidin or acetylstrophanthidin of $10^{-5}$M or greater appear to maximally inhibit the $Na^+$-$K^+$ pump and produce a rapid rise of $aNa_i$. The $aNa_i$ does not continue to rise, however, but slows drastically within about 30 min and $aNa_i$ tends to approach a plateau level of about 20-40 mM (Fig. 1 and Fig. 7, reference 4). It would appear that other mechanisms are able to maintain $aNa_i$ at relatively low levels when the $Na^+$-$K^+$ pump is blocked. We concluded (3) that this was primarily due to an exchange of $Na_i^+$ for $Ca^{2+}$ across the cell membrane. Raising the $[Ca]_o$ reduced $aNa_i$ while lowering $[Ca]_o$ increased $aNa_i$. These changes occurred in the presence of high concentrations of cardioactive steroids and so were not mediated via the action of the $Na^+$-$K^+$ pump. Strontium and $Ca^{++}$ were interchangeable in this process but $Mg^{2+}$ and $Mn^{2+}$ were not. In fact $Mn^{2+}$ inhibited the ability of high $[Ca]_o$ to decrease $aNa_i$.

What appears to be a $Na^+$-$H^+$ exchange across the cell membrane has been demonstrated in snail neurones (16). A $Na_i^+$-$H_o^+$ exchange might help to maintain $aNa_i$ at relatively low plateau levels in high cardioactive steroid concentrations in cardiac tissues. If such an exchange existed then we would anticipate an intracellular acidification under $Na^+$-loading conditions. Purkinje fibres do acidify in the presence of high cardioactive steroid concentrations as is illustrated in Fig. 1. During the first few minutes of exposure to strophanthidin, when $aNa_i$ was rising rapidly, there was no change in $pH_i$, but as $aNa_i$ began to approach a plateau level the $pH_i$ started to decrease. Presumably $pH_i$ regulatory mechanisms prevented an even larger decrease of $pH_i$ which tended to stabilize at a new, more acidic, level.

An alternative explanation for these results is that the acidification was caused as a result of intracellular $Ca^{2+}$-loading in the same way that $Ca^{2+}$ injection into snail neurones causes large intracellular acidifications (10). High levels of internal $Ca^{2+}$ might produce this effect by a number of mechanisms e.g. by causing $H^+$ efflux from mitochondria as a $Ca^{2+}$-$H^+$ exchange occurs at the mitochondrial membrane (1, 18), by $Ca^{2+}$-$H^+$ exchange at the cell membrane, or by a simple displacement of $H^+$ from internal binding sites (analogous to the $Ca^{2+}$-$H^+$-$EGTA^4$-buffer interactions).

Reduction of $[Na]_o$ results in an increased $Ca^{2+}$ content of cardiac muscle (11) presumably due to a $Na_i^+$-$Ca_o^{2+}$ exchange (9, 12). A decrease in $[Na]_o$ causes a large decrease in $aNa_i$ in sheep heart Purkinje fibres that is independent of the activity of the $Na^+$-$K_+$ pump and is inhibited, but not completely blocked by reduction or removal of $[Ca]_o$. Fig. 2 shows an experiment where the effect on $aNa_i$ and $pH_i$ of a reduction of $[Na]_o$ to one tenth normal is illustrated.

Any changes in $pH_i$ in the absence of strophanthidin were found to be small. Following the addition of a high concentration of the cardioactive steroid (Fig. 2, right) a reduction of $[Na]_o$ still resulted in a very large decrease in $aNa_i$ but now a very large intracellular acidification occurred. This might be explained in terms of a $Na^+$-$H^+$ exchange with a larger decrease of $aNa_i$ in the presence of the cardioactive steroid (see also reference 5) but perhaps a more likely explanation would be an effect on $pH_i$ via a change in $aCa_i$. The internal $Ca^{2+}$ would be probably have been high in the presence of the cardioactive steroid. This could have produced a moderate to severe $Ca^{2+}$ loading where the $Ca^{2+}$ uptake systems would have been working at a high rate to maintain $aCa_i$ levels low. Under these conditions a reduction of $[Na]_o$ would produce an extra large $Ca^{2+}$ influx due to the increased $aNa_i$ (9).This large $Ca^{2+}$ influx into cells already $Ca^{2+}$-loaded

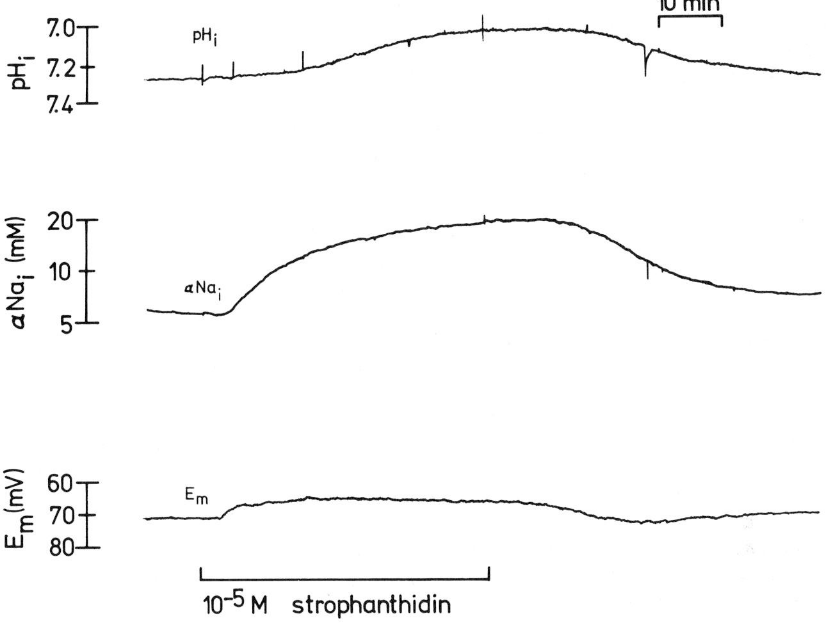



Fig. 1. Simultaneous measurements of the $pH_i$ (upper trace), $a^i_{Na}$ (middle trace) and $E_m$ (lower trace) in the same Purkinje fibre. $10^{-5}$M strophanthidin was added for about 45 min. The $E_m$ electrode served as the reference for both the $Na^+$ and the pH-selective microelectrodes

could account for the large intracellular acidification observed under these conditions. The size of the acidification shown in Fig. 2 represents a large change in $H^+$ ions. If the intracellular buffering power of sheep heart Purkinje fibres under these conditions is taken as 35 mequiv. $H^+$/pH unit/litre (7) then the acidification is indicative of an increase in sarcoplasmic $H^+$ ions of approximately 16 mequiv./litre. This is of the same order of magnitude as the decrease of $aNa_i$. If the intracellular $Na^+$ activity coefficient is assumed to be the same as that of the extracellular solution, then the ratio of the change of $H^+$ ions was 1.6:1.

If the change in $pH_i$ was produced as a result of $Ca^{2+}$ entry into the cells then a comparable change in total internal $Ca^{2+}$ would be expected. However, as intracellular $Ca^{2+}$ is heavily buffered (e.g. by sarcoplasmic reticulum, mitochondria, cellular membranes and proteins) the change in $aCa_i$ may well be expected to be small. From the results it is clear that we needed to measure changes of intracellular $Ca^{2+}$ under these conditions in order to assess the validity of our explanation of the results. Such an experiment is illustrated in Fig. 3.

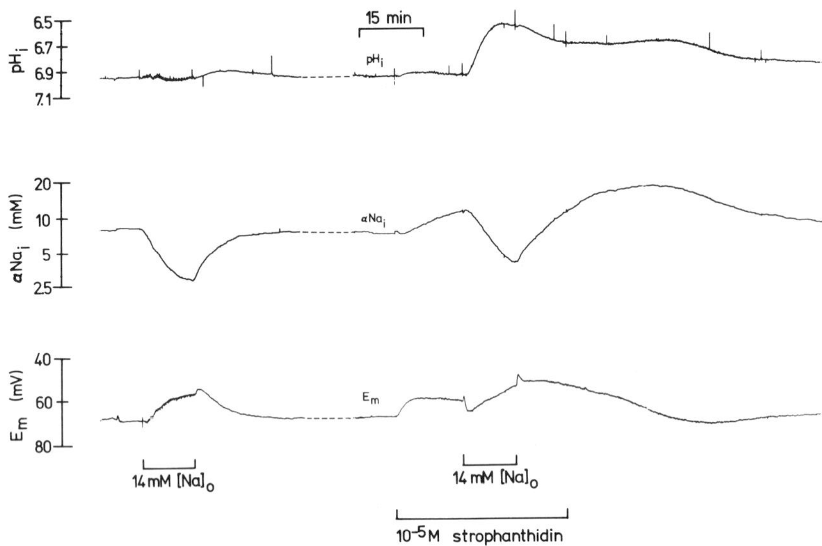



Fig. 2. Effects on the $pH_i$ and the $aNa_i$ of reducing the $[Na]_o$ to one tenth normal in the absence and in the presence of $10^{-5}$M strophanthidin. The dashed line indicates a break in the recording for a period of 45 min.

We have used two parameters to assess changes in $aCa_i$ (1) measurement of the force of contraction (contracture) of the muscle and (2) a direct estimate using $Ca^{2+}$-selective microelectrodes. We have found that a reduction of $[Na]_o$ to one tenth normal usually produces a small contracture in sheep heart Purkinje fibres. In the experiment illustrated no increase in tension is apparent (the small change in baseline tension and the blips at the time of reduction, and return to $[Na]_o$ are solution flow artifacts). A small increase in $aCa_i$ is detectable with the $Ca^{2+}$-selective microelectrode.

Following addition of 2.5 x $10^{-5}$ M acetylstrophanthidin there was an almost immediate rise in $aNa_i$. This was followed, after a delay, by a small slow increase in $aCa_i$. When $[Na]_o$ was now reduced to one tenth normal, in the presence of the cardioactive steroid, there was a large reduction in $aNa_i$ accompanied by a large increase in $aCa_i$. This increase in $aCa_i$ resulted in the development of a large contracture which relaxed spontaneously. The $aCa_i$ in the cell penetrated also decreased but not as rapidly as the contracture tension of the preparation as a whole. It is possible that part of the decline in contracture tension is due to the acidification demonstrated in Fig. 2, as acidosis depresses contractility (e.g. 8). Return to a normal $[Na]_o$, still in the presence of the cardioactive steroid, resulted in the return of $aNa_i$ to a high level, and a decrease in $aCa_i$ and contracture tension. Continued exposure to acetylstrophanthidin produced a slow further increase in $aNa_i$ accompanied by an increase in $aCa_i$ and tension. Results from experiments of this type are therefore consistent with the idea that the large $pH_i$ change (Fig. 2) on reduction of $[Na]_o$ in the presence of cardioactive steroids is $Ca^{2+}$ mediated.

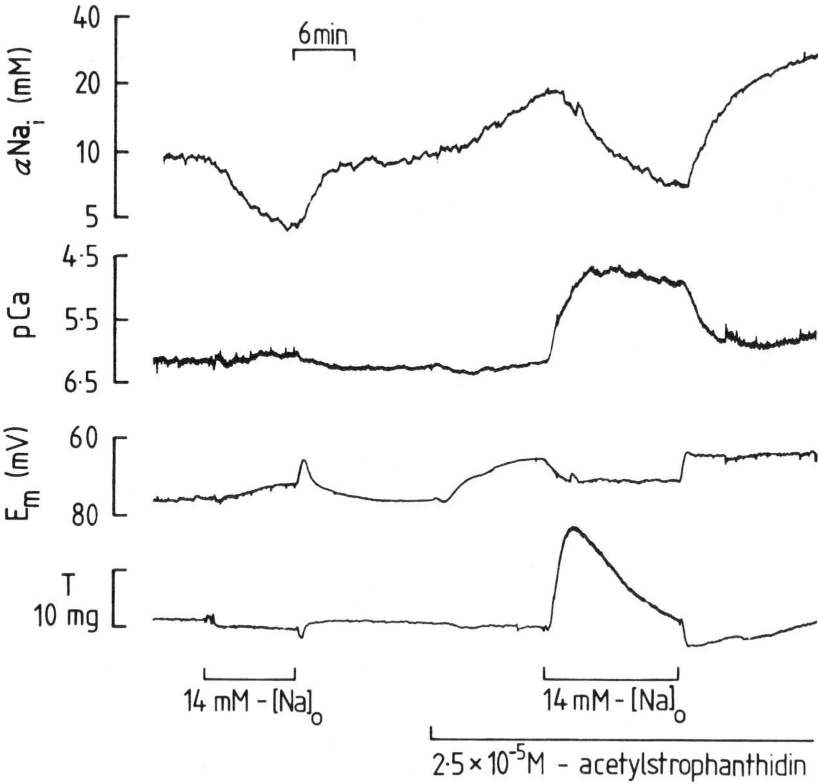



Fig. 3.  Pen recording of an experiment showing the effects on: intracellular $Na^+$ ($aNa_i$) and $Ca^{2+}$ ($aCa_i$) activities, membrane potential ($E_m$) and tension (T) with a reduction of $[Na]_o$ (replaced by Tris) in the absence and presence of $2.5 \times 10^{-5}M$ acetylstrophanthidin.

## Conclusions

In this work we have used a variety of ion-selective microelectrodes to monitor $Na^+$, $Ca^{2+}$ and $H^+$ ions in cardiac cells. We have also measured the tension produced by the preparations as an independent measure of $Ca^{2+}$ levels in the cells. The results have indicated the need for this type of approach as changes in the activity of one type of ion in the cell can alter the activity of other ionic species. This situation is not unexpected when different ions share the same intracellular buffering systems (e.g. $Ca^{2+}$ and $H^+$ interaction at mitochondria binding sites).

We have shown that in the presence of cardioactive steroids, changes in $[Na]_o$ can produce large changes in $aNa_i$, $aCa_i$ and $pH_i$. It seems likely that these changes originate from alteration in the $Na^+-Ca^{2+}$ exchange system at the cell membrane. High $aCa_i$ levels and intracellular $Ca^{2+}$ loading are aggravated by reduction of $[Na]_o$. The resulting large increase in $aCa_i$ could then be responsible for the large decrease in $pH_i$. It is to be expected that such large changes in $pH_i$ and $aCa_i$ also affect many other systems in the cell.

Clearly the improvement of ion-selective microelectrodes for assaying as many of the physiologically interesting ions in the cell as possible should allow a more accurate picture to be formed of ion regulating mechanisms.


## References


1. Bartley W. Amoore JE (1958) The effects of manganese on the solute content of rat liver mitochondria. Biochem J 69: 348
2. Deitmer JW, Ellis D (1978a) Changes in the intracellular sodium activity of sheep heart Purkinje fibres produced by calcium and other divalent cations. J Physiol 277: 437
3. Deitmer JW, Ellis D (1978b) The intracellular sodium activity of cardiac Purkinje fibres during inhibition and reactivation of the Na-K pump. J Physiol 284: 241
4. Deitmer JW, Ellis D (1980a) The intracellular sodium activity of sheep heart Purkinje fibres: Effects of local anaesthetic and tetrodotoxin. J Physiol 300: 269
5. Deitmer JW, Ellis D (1980b) Interactions between the regulation of the intracellular pH and sodium activity of sheep cardiac Purkinje fibres. J Physiol 304: 471
6. Ellis D (1977) The effect of external cations and ouabain on the sodium activity in sheep heart Purkinje fibres. J Physiol 273: 211
7. Ellis D, Thomas RC (1976) Direct measurement of the intracellular pH of mammalian cardiac tissue. J Physiol 262: 755
8. Fabiato A, Fabiato F (1978) Effects of pH on the myofilaments and the sarcoplasmic reticulum of skinned cells from cardiac and skeletal muscle. J Physiol 276: 233
9. Glitsch HG, Reuter H, Scholz H (1970) The effect of the internal sodium concentration on calcium fluxes in isolated guinea-pig auricles. J Physiol 209: 25
10. Meech RW, Thomas RC (1977) The effect of calcium injection on the intracellular sodium and pH of snail neurones.
11. Niedergerke R (11963) Movements of Ca in  frog ventricles at rest and during contractures. J Physiol 167: 515
12. Reuter H, Seitz N (1968) The dependence of calcium efflux from cardiac muscle on temperature and external ion composition. J Physiol 195: 451
13. Oehme M, Kessler M, Simon W (1976) Neutral carrier $Ca^{2+}$-microelectrode. Chimica 30: 204
14. Thomas RC (1970) New design for sodium sensitive glass microelectrode. J Physiol 210: 82
15. Thomas RC (1974) Intracellular pH of snail neurones measured with a new pH-sensitive glass microelectrode. J Physiol 238: 159
16. Thomas RC (1977) The role of bicarbonate, chloride and sodium ions in the regulation of intracellular pH in snail neurones. J Physiol 273: 317
17. Thomas RC (1978) Ion-sensitive intracellular microelectrodes: How to make and use them. Academic Press, London-New York-San Francisco

18. Vercesi A, Reynafarje B, Lehringer Al (1978) Stoichiometry of $H^+$ ejection and $Ca^{2+}$ uptake coupled to electron transport in rat heart mitochondria. J Biol Chem 253: 6379
19. Weingart R (1977) The actions of ouabain on intracellular coupling and conduction velocity in mammalian ventricular muscle. J Physiol 264: 341


## Acknowledgements

D.E. wishes to thank the MRC, and J.W.D. the D.F.G. for support. D.M.B. is an American Heart Association (G.L.A. Affiliate) Fellow. The earlier experiments described in this work were done at the Physiology Department, University of Bristol on an MRC grant to Dr. R.C. Thomas.


Department of Physiology, Medical School, Teviot Place, Edinburgh EH8 9AG, Scotland.
* Present address: Abteilung Biologie, Ruhr-Universität, D-4630 Bochum 1, FRG
** Present address: Department of Physiology, UCLA School of Medicine, Los Angeles, Ca 90024, U.S.A.

# The Electrogenic Na-K Pump in the Sheep Cardiac Purkinje Fibre

R.D. VAUGHAN-JONES, W.J. LEDERER[*], D.A. EISNER[**]

## Introduction

It is known that the Na-K pump in a wide variety of tissue can be electro-genic (see 13 for a review). The experiments described here examine the electrogenic Na-K pump in the sheep cardiac Purkinje fibre under conditions where the pump is stimulated by an elevation of internal Na. Recent reports have suggested that it is possible to measure directly the electrogenic Na pump current in this tissue using the voltage clamp technique (5, 7). Such measurements were originally made by Thomas (12) in snail neurones. He stimulated the Na-K pump with an intracellular iontophoretic injection of Na and measured an outward current under voltage-clamp conditions. By simultaneously monitoring the intracellular Na activity, $a_{Na}^i$, using a glass $Na^+$-sensitive microelectrode he observed that as the excess Na was pumped out of the cell both the outward current and $a_{Na}^i$ declined expo-nentially with the same time-constant confirming that the current was produced by activation of the Na-K pump.

It is now possible to perform essentially similar experiments using the sheep cardiac Purkinje fibre (6). We have simultaneously measured membrane current and $a_{Na}^i$ during stimulation of the Na-K pump. In this way we can examine directly the relationship between the electrogenic Na pump current and $a_{Na}^i$. Furthermore, we have also recorded the twitch and tonic tension of the fibre thus permitting tension changes to be compared with changes of $a_{Na}^i$.

## Methods

Recessed-tip $Na^+$-sensitive glass microelectrodes were used for the intra-cellular measurements of $a_{Na}^i$. They were constructed conventionally (see 14) with tips of $<$ 1.0/um. They responded to a fast change of [Na] with a 50% response time $(t_{0.5})$ of $<$ 6 sec. This was required since many of the changes of $a_{Na}^i$ to be measured occurred with half-times of 30 - 70 sec. The electrodes were calibrated using NaCl solutions of varying [Na] (substi-tuting KCl for NaCl; [KCl] + [NaCl] = 145 mM), and were considered accept-able if they produced 53 - 61 mV response for a ten-fold change in Na ac-tivity (activity coefficient at $36^O$C = 0.75). In general a fresh electrode was used for each experiment.

Purkinje fibres were dissected from fresh sheep hearts, a suitable fibre being cut to about 3 mm in length. One end was then pinned in the experi-mental bath whilst the other was connected to a piezoresistive tension transducer (4). A two-microelectrode voltage clamp technique was used to control membrane potential. The $a_{Na}^i$ was monitored simultaneously as the difference between the voltage-recording and $Na^+$-sensitive microelectrode when these were intracellular. All microelectrodes were positioned in a line about 50/um to 100/um apart (order: current, voltage, $Na^+$).

Normal cardiac Tyrode contained 145 mM NaCl; 4 mM KCl; 2 mM $CaCl_2$; 1 mM $MgCl_2$; 10 mM Tris HCl; 10 mM glucose. Modifications to this are indicated in the text. For example, a solution described as 4 mM $[Rb]_o$ is the above solution with KCl removed and 4 mM RbCl added. The pH of all solutions was $7.4 \pm 0.1$ and temperature was $35 - 37°C \pm 0.5°C$.


## Results

Fig. 1 shows an experiment where $a_{Na}^i$, membrane current and tension were recorded simultaneously. In this experiment all solutions were $K^+$-free and contained varying amounts of $Rb^+$. We find that $K^+$ and $Rb^+$ activate the Na-K pump about equally and since it has already been shown (5) that the electrogenic $Na^+$ pump current can be measured more accurately in $Rb^+$-containing rather than $K^+$-containing solutions, Rb has also been used in subsequent experiments. The preparation was initially bathed in 10 mM $[Rb]_o$ and the membrane potential was held by the voltage-clamp circuit at -68 mV. Every 10 sec a depolarising step to -35 mV was applied for 500 msec (shown by the upward deflections on the voltage trace). This elicited a twitch which was recorded as a small spike on the tension trace. The rapid current injections required to produce a twitch were too brief to be recorded on the current trace which therefore shows only the slower changes of membrane current at the holding potential.

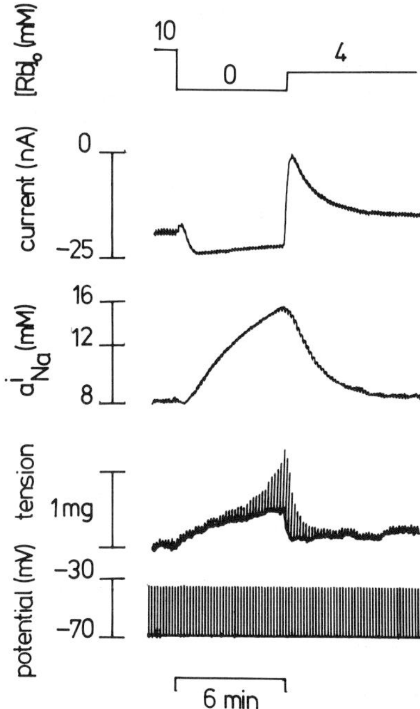


Fig. 1. Simultaneous measurement of membrane current, $a_{Na}^i$, tension and voltage clamped membrane potential. The current and $a_{Na}^i$ records have been filtered; time constant 3 sec (from Eisner, Lederer & Vaughan-Jones, 1980 (6))

Removing external $Rb^+$ largely inhibited the Na-K pump producing a rise of $a_{Na}^i$. This was accompanied by an increase of both twitch and tonic tension. Upon re-adding 4 mM $[Rb]_o$, $a_{Na_i}$ and both twitch and tonic tension recovered close to control levels. As $a_{Na}$ declined there was a transient overshoot of membrane current which then also declined. This transient is outward relative to the final holding current and has been ascribed to the electrogenic extrusion of $Na^+$ (5, 7). It is thought to reflect the recovery of $a_{Na}^i$. Fig. 2 shows the recovery of $a_{Na}^i$ (open circles) and of membrane current (filled circles), following reactivation of the Na-K pump, plotted on semi-logarithmic axes (data from Fig. 1). The level of $a_{Na}^i$ and current above the final resting levels ($\Delta a_{Na}^i$, and $\Delta I$) are shown on the ordinates. After an initial period during which the membrane current transient was rising to its peak value, both $\Delta a_{Na}^i$ and $\Delta I$ decline exponentially with virtually the same half-time (48 sec and 49 sec respectively). This confirms that the fall of the outward current transient indeed reflects the fall of $a_{Na}^i$ upon reactivating the Na-K pump with external $Rb^+$. This in turn is powerful evidence that the outward current transient represents electrogenic $Na^+$ extrusion. Furthermore, if the fall of $a_{Na}^i$ represents the net extrusion of $Na^+$ by the Na-K pump, then the parallel, exponential recovery of both current and $a_{Na}^i$ means that a constant fraction of $Na^+$ extrusion is electrogenic. In other words, the coupling of $Na^+$ efflux to, in this case, $Rb^+$ uptake is constant during the recovery. The open triangles in Fig. 2 show the recovery of $a_{Na}^i$ in the same experiment if 4 mM $[K]_o$ instead of 4 mM $[Rb]_o$ is used to reactivate the Na-K pump. It is clear that $a_{Na}^i$ re-

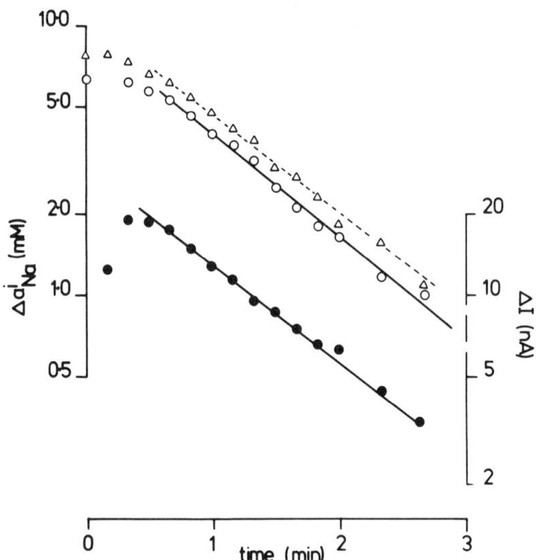

Fig. 2. Fall of $a_{Na}^i$ and outward current transient upon re-activation of Na-K pump with 4 mM $[Rb]_o$. Values of $a_{Na}^i$ (open circles) and current (filled circles) above their final level ($\Delta a_{Na}^i$, $\Delta I$) are shown on ordinate (log scale). Data from Fig. 1. Open triangles show a similar recovery of $a_{Na}^i$, but using 4 mM $[K]_o$. Lines of best fit, ignoring first 20 sec after re-adding $[Rb]_o$

covers at virtually the same rate, implying that both cations are essentially equivalent in their ability to activate the pump. $Rb^+$ is therefore a good substitue for $K^+$ in these experiments.

## The Dependence of Electrogenic $Na^+$ Extrusion upon $a_{Na}^i$

In order to explore further the relationship between $a_{Na}^i$ and the electrogenic $Na^+$ pump current, a Purkinje fibre was exposed to $Rb^+$-free, $K^+$-free solution for different durations in order to achieve different levels of $a_{Na}^i$ (two exposures are shown in Fig. 3A). After raising $a_{Na}^i$, the Na-K pump was then re-activated with 10 mM $[Rb]_o$. In each case this produced a recovery of $a_{Na}^i$ accompanied by the rise and fall of an outward current transient. The half-times of recovery of $a_{Na}^i$ and current following four different periods of exposure to $K^+$-free, $Rb^+$-free solution are listed in Table 1. In each case, both current and $a_{Na}^i$ declined exponentially in parallel. Furthermore all four re-activations occurred with similar half-times. From this and other experiments it is not possible to say if the small variations of $t_{0.5}$ during different re-activations are significant. In general, over the range of $a_{Na}^i$ tested (up to 16.5 mM), the rate of recovery of $a_{Na}^i$ and current and hence of electrogenic Na-K pumping is linearly related to $a_{Na}^i$. There is no sign of saturation of the recovery rate as $a_{Na}^i$ is elevated.

Table 1. Half-times of recovery of $a_{Na}^i$ ($t_{0.5}^{Na}$) and outward current transient ($t_{0.5}^I$) after different periods of $Na^+$-loading in $K^+$- free solution ($\Delta t$) (Continuous recording from a single fibre).

| t min | $t_{0.5}^{Na}$ | $t_{0.5}^I$ | $t_{0.5}^{Na} / t_{0.5}^I$ |
|---|---|---|---|
| 1.5 | 36.6 | 37.7 | 0.97 |
| 2.5 | 33 | 35.4 | 0.93 |
| 5 | 33.4 | 34.2 | 0.98 |
| 10 | 30.2 | 29.4 | 1.03 |

In order to see if the electrogenic fraction of $Na^+$ extrusion changes as $a_{Na}^i$ is varied, reference can again be made to Fig. 3A. During any reactivation the area under each outward current transient (i.e. the time integral of the outward current) is proportional to the amount of charge extruded electrogenically by the Na-K pump ($Q^e$). The corresponding change of $a_{Na}^i$ ($\Delta a_{Na}^i$ during the recovery is proportional to the total amount of $Na^+$ extruded. The ratio between these two measurements is thus proportional to the electrogenic fraction of the $Na^+$ extrusion. Fig. 3B shows that these two parameters ($\Delta a_{Na}^i$ and $Q^e$) obtained from a continuation of the experiment shown in Fig. 3A, increase proportionately as would be expected for a constant ratio. In three fibres there was no evidence that this ratio changed significantly in any systematic way. This is in agreement with the fact that during any reactivation both $a_{Na}^i$ and the electrogenic current decline exponentially in parallel. Moreover it is consistent with a fixed coupling ratio for the Na-K pump which is unaffected by changes of $a_{Na}^i$. An estimate of the exact fraction of electrogenic $Na^+$ extrusion during a reactivation can be made if the intracellular volume of the Purkinje fibre is known, assuming also that the intracellular activity coefficient for $Na^+$ at $36^oC$ is 0.75 (as in external Tyrode). A rough measure of volume was made

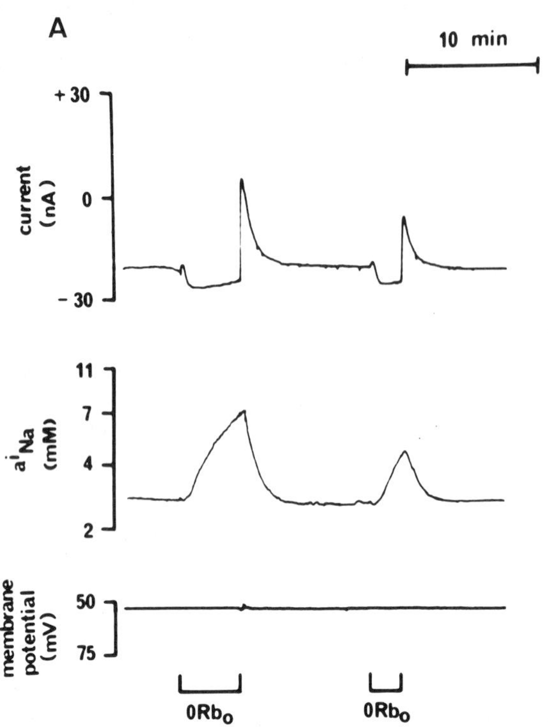



Fig. 3A.
Effect of two different durations of exposure (5 and 2.5 min) to $0 \, [Rb]_O$ (all solutions were $K^+$-free). Control solution contained 10 mM $[Rb]_O$

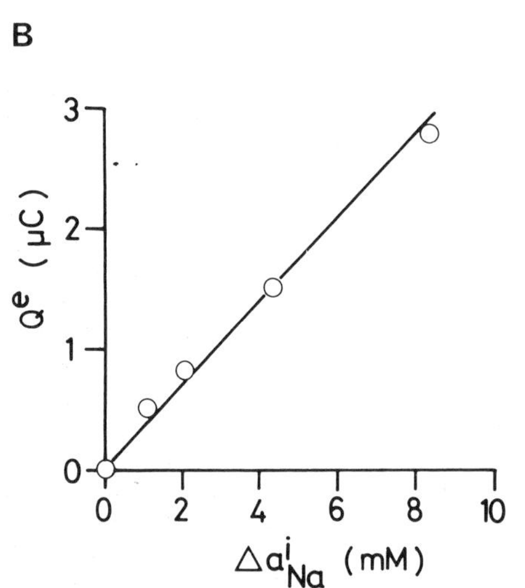



Fig. 3B.
Plot of area under outward current transient (expressed in $\mu$ Coulombs) versus change of $a^i_{Na}$ ($\Delta a^i_{Na}$) upon re-activation with 10 mM $[Rb]_O$ following 4 different durations of exposure to $0 \, [Rb]_O$ (same preparation as in A)

during each experiment using a binocular microscope (core-diameter and length). Clearly this can at best be only approximate. The mean value of the electrogenic fraction from 8 preparations (30 re-activations) was 0.26 $\pm$ 0.06 (SEM). This indicates that, on average, 26% of the Na$^+$ extrusion appeared to be electrogenic.


## Discussion


The electrogenic Na-K pump in the cardiac Purkinje fibre.

When the Na-K pump is substantially inhibited by removing the external activator cation (in this case K$^+$ or Rb$^+$) then $a_{Na}^i$ rises. Upon re-activating the pump by adding back K$^+$ or Rb$^+$, there is a net efflux of Na$^+$ from the cell stimulated by the raised level of $a_{Na}^i$. Whilst using external Rb$^+$ we have shown that re-activation of the Na-K pump produced an exponential decline of $a_{Na}^i$ in agreement with the results of Deitmer & Ellis (2). This suggested that the rate of Na-K pumping is linerarly related to $a_{Na}^i$ as also suggested for the snail neurone (12,13) and the squid axon (9). In Purkinje fibres this relationship appears linear for quite a large range of $a_{Na}^i$ (1), although we cannot of course exclude a more complex relationship either at very high or very low $a_{Na}^i$. The recovery of $a_{Na}^i$ is paralleled by the appearance of an outward current which relaxes exponentially with virtually the same time-constant. Hence this current is lineraly related to $a_{Na}^i$. Both the outward current transient and the recovery of $a_{Na}^i$ become larger if the preloading with internal Na$^+$ is prolonged, and although not reported here, both the rate constant of recovery of $a_{Na}^i$ and the current transient are equally sensitive to $[Rb]_o$ (unpublished results). Furthermore the outward current transient is known to be abolished by cardioactive steroids (5,7) and so is the recovery of $a_{Na}^i$ (unpublished results). Consequently our results greatly strengthen the argument that the outward current transient represents electrogenic Na$^+$ extrusion by the Na-K pump. In addition they are in agreement with those of Thomas (12,13) who demonstrated the same parallel behaviour of membrane current and $a_{Na}^i$ in a snail neurone following an intracellular injection of Na$^+$.

The parallel behaviour during a re-activation also means that a constant fraction of Na$^+$ is being extruded electrogenically. If this is achieved by means of the Na-K pump (in our case when re-activated by external Rb$^+$ which is equivalent to external K$^+$) then the counter coupling of Na$^+$ to Rb$^+$ transport must also be constant. This is further supported by the fact that the total charge transfer appears to be a constant fraction of the total change of $a_{Na}^i$ and is independent of variation of $a_{Na}^i$. Further experiments indicate that it is also independent of $[Rb]_o$ from 1-20 mM (unpublished results). Our results are therefore consistent with the idea that the coupling ratio of the Na-K pump is fixed, at least for the ranges of $a_{Na}^i$ and $[Rb]_o$ tested in our experiments. The same conclusion was reached by Thomas (12) for snail neurones but apart from this there is as yet no general agreement concerning the constancy or otherwise of the Na-K pump's stoichiometry in various excitable cells.

It is also important to consider the absolute value for the stoichiometry of the Na-K pump. Our experiments suggest that, on average 26% of the extrusion of Na$^+$ during a re-activation is electrogenic. Again, if the rest of the Na$^+$ extrusion was balanced by an uptake of Rb$^+$, then the stoichiometry would be about $4Na^+ : 3Rb^+$. This is close to the values estimated by Thomas (12) in snail (between $4Na^+ : 3K^+$ and $3Na^+ : 2K^+$) and close to the $3Na^+ : 2K^+$ traditionally quoted for the red blood cell (e.g. 8). However, we are at present cautious about our own value since it relies heavily upon intracellular volume measurements which are difficult to obtain accurately. Nevertheless the apparent stoichiometry in heart Purkinje fibres does not appear greatly different from that usually quoted in other tissues.

One final point must be considered, and that is the possible participation in our experiments of $Na^+$ transport mechanisms other than the Na-K pump. Evidence is accumulating for $Na^+-Ca^{++}$ and $Na^+-H^+$ exchange in cardiac Purkinje fibres (e.g. 2, 3); consequently, will these systems distort our assessment of the Na-K pump? There is no fool-proof answer to this. But, even if other $Na^+$-systems were operating, the clearly linear relation between $a_{Na}^i$ and the electrogenic Na-K pump current, as well as their exponential decline during a re-activation would be extremely difficult to explain unless (i) $Na^+-K^+$ pumping was linearly related to $a_{Na}^i$ and (ii) the coupling ratio of the $Na^+-K^+$ pump was constant. Less certain perhaps would be our estimate of the absolute stoichiometry ($4Na^+ : 3Rb^+$). Therefore until more is known about $Na^+-H^+$ and $Na^+-Ca^{++}$ exchange in heart muscle the exact coupling coefficient may remain unresolved.

## Relationship between $a_{Na}^i$ and Tension

Because we simultaneously measured tension, we are able to relate changes of $a_{Na}^i$ caused by inhibition and re-activation of the Na-K pump with changes of both twitch and tonic tension. Our experiments provide a direct demonstration that the rise of both twitch and tonic tension upon inhibition of the Na-K pump occurs at a rate comparable to the rise of $a_{Na}^i$. This agrees well with the idea that the positive inotropic effect may be mediated via changes of $a_{Na}^i$ which secondarily alter $a_{Ca}^i$ possibly by means of an $Na^+/Ca^{++}$ exchange mechanism (e.g. 10). A similar relation between twitch tension and $a_{Na}^i$ has been reported by Lee et al. (11) during Na-K pump inhibition with dihydrouabain. Of course, a simple scheme whereby an exchange of $Na^+$ for $Ca^{++}$ across the sarcolemma accounts for the positive inotropic effect may not be wholly adequate. In this respect, it is interesting to note that tonic tension frequently recovers very rapidly upon re-activation of the Na-K pump, whereas the twitch usually recovers more slowly as does $a_{Na}^i$ (see Fig. 1). This suggests that the relationship between $a_{Na}^i$ and twitch and tonic tension is rather complex. It is clear nevertheless that a rise and fall of $a_{Na}^i$ in cardiac Purkinje fibres is associated with positive and negative inotropic effects.

## References


1. Deitmer JW, Ellis D (1978a) The intracellular sodium activity of cardiac Purkinje fibres during inhibition and re-activation of the Na-K pump. J Physiol 284: 241
2. Deitmer JW, Ellis D (1978b) Changes in the intracellular sodium activity of sheep heart Purkinje fibres produced by calcium and other divalent cations. J Physiol 277: 437
3. Deitmer JW, Ellis D (1980) Interactions between the regulation of the intracellular pH and sodium activity of sheep cardiac Purkinje fibres. J Physiol 304: in press

4. Eisner DA, Lederer WJ (1979) Inotropic and arrhythmogenic effects of potassium-depleted solutions on mammalian cardiac muscle. J Physiol 294: 255

5. Eisner DA, Lederer WJ (1980) Characterisation of the electrogenic sodium pump in cardiac Purkinje fibres. J Physiol 303: 441

6. Eisner DA, Lederer WJ, Vaughan-Jones RD (1980) Electrogenic sodium pumping in cardiac muscle: simultaneous measurement of intracellular sodium activity, membrane current and tension. J Physiol 300: 42

7. Gadsby DC, Cranefield PF (1979) Direct measurement of changes in sodium pump current in canine cardiac Purkinje fibres. Proc Natl Acad Sci USA 76: 1783

8. Garrahan PJ, Glynn IM (1967) The stoichiometry of the sodium pump. J Physiol 192: 217

9. Hodgkin Al, Keynes RD (1956) Experiments on the injection of substances into squid giant axons by means of a microsyringe. J Physiol 131: 592

10. Langer GA (1970) The role of sodium ions in the regulation of myocardial contractility. J mol cell Cardiol 1: 203

11. Lee CO, Kang DH, Sokol JH, Lee KS (1979) Relation between intracellular Na activity and positive inotropic action of sheep cardiac Purkinje fibres. Biophys J 25: 198a

12. Thomas RC (1969) Membrane current and intracellular sodium changes in a snail neurone during extrusion of injected sodium. J Physiol 201: 495

13. Thomas RC (1972) Electrogenic sodium pump in nerve and muscle cells. Physiol Rev 52: 563

14. Thomas RC (1978) Ion sensitive intracellular microelectrodes. How to make them and how to use them. Academic Press, Biol Techniques Series

University of Oxford, Dept. of Pharmacology, South Parks Road, Oxford OX1 3QT/GB

*University of Maryland, Department of Physiology, Baltimore

**Physiological Laboratory, Cambridge

# Measurements of Myocardial Extracellular Na⁺, K⁺, Ca²⁺, and H⁺ Using Ion-Selective Electrodes During Ischemia

H. HIRCHE, R. BISSIG, R. FRIEDRICH, U. KEBBEL, V. ZYLKA

It is well known that during ischemia $Na^+$, $Ca^{2+}$, and $H_2O$ diffuse into the myocardial cell while $K^+$ and $H^+$ are released. However, very little information is available about the rate and the time course of these transmembrane ion- and $H_2O$-fluxes during ischemia. It was the aim of this study to investigate these transmembrane fluxes by measuring the extracellular cation-concentrations and the extracellular osmolality of the ischemic myocardium.

## Methods

The type of ion-selective membrane electrodes which were used to measure the extracellular concentrations of $Na^+$, $K^+$, $Ca^{2+}$ concentrations of the myocardium ($[Na^+]e$, $[K^+]e$, $[Ca^{2+}]e$) are shown in Fig. 1. In the majority of

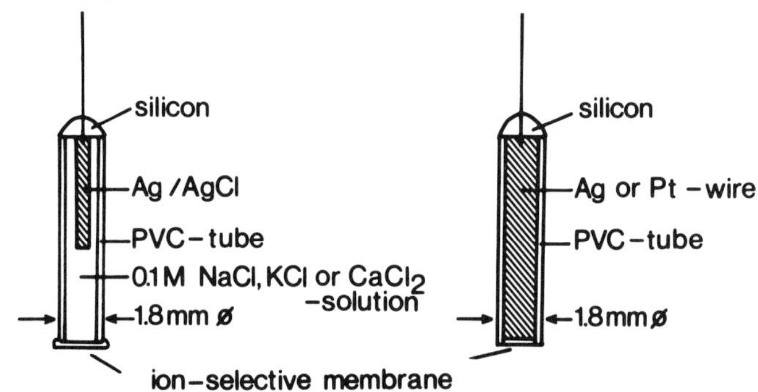


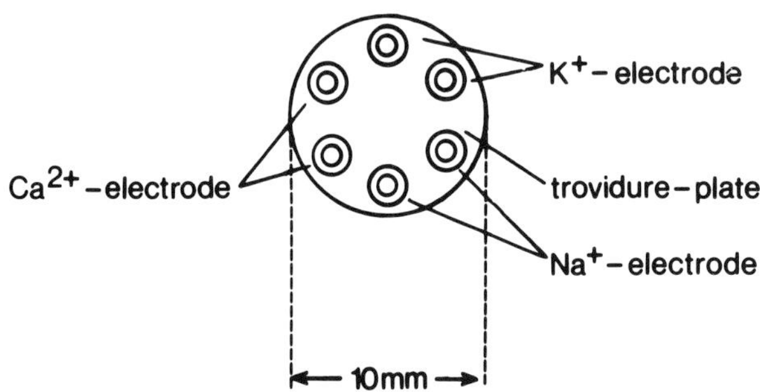


Fig. 1. Schematic drawing of ion-selective liquid membrane electrodes with electrolyte bridge (left side top) and solid-contacted electrodes (right side top) according to Höper et al. (7). These electrodes (outer diameter 1.8 mm) were stuck into the corresponding holes of a trovidure-plate which was sewn onto the epicardium using three atraumatic threads. For details see text

experiments ion-selective electrodes with an electrolyte bridge between the ion-selective membrane and the Ag/AgCl electrode were used. Solid contacted electrodes (Fig. 1) which were at first described by Höper et al. (7) were mainly used for the construction of catheter electrodes to measure ion activities in blood vessels. The ion-selective membranes were composed as shown in table 1 according to W. Simon (personal communication). The trovidure plates with up to 6 ion-selective electrodes were sewn on the epicardium of the left ventricle with three atraumatic threads within the centre of the ischemic area.

Table 1.  Composition of ion-selective membranes

| | |
|---|---|
| $K^+$ | 1.3 wt.-% valinomycin<br>69.1 wt.-% DOS (dioctyl-sebacic acid)<br>29.6 wt.-% PVC |
| $Na^+$ | 3 wt.-% carrier*<br>64 wt.-% DOS<br>33 wt.-% PVC |
| $Ca^{2+}$ | 1 wt.-% carrier*<br>66 wt.-% o-NPOE (o-nitro-phenyloctyl-ether)<br>33 wt.-% PVC |

*neutral carriers according to Ammann et al. (1)

Extracellular $H^+$ activity of the myocardium ( $[H^+]e$, pHe) was measured with bulb-type buffer-filled glass minielectrodes according to Gebert et al. (2). These electrodes made of Corning glass 0150 had a tip diameter of about 50 /um and were pushed 5 - 6 mm deep into the left ventricular wall as close as possible to the ion-selective electrodes within the ischemic area. As reference electrode for all ion-selective electrodes one conventional calomel electrode was used which was inserted several cm deep between thoracic muscles.

In three pigs $[K^+]e$ was also measured using mini-electrodes as described recently by Hill et al. (3, 4). These electrodes were inserted as close as possible to our surface electrodes in the centre of the ischemic region about 3 - 4 mm deep into the myocardium.

In two different pigs another reference method to measure $[K^+]e$ was used. Extracellular fluid was absorbed with thin filter paper discs as described below for determination of extracellular osmolality. 80 /ul of fluid was absorbed in a filter paper disc with a diameter of 10 mm. Then an ion-selective electrode as shown in Fig. 1 with a small reference electrode was placed on the filter paper disc. The measurements were performed in a small plexi-glass container to prevent evaporation of the fluid which was absorbed in the filter paper. Some preliminary results are shown in Fig. 2.

The extracellular osmolality was measured with a vapor pressure Wescor osmometer. 8 /ul of solution are necessary for one determination. Samples of the extracellular fluid of the ischemic myocardium were absorbed with small filter paper discs (6 mm diameter) from the epicardium. The epicardium was prevented from drying and evaporation by an insulating cover of thin rubber, under which the filter paper discs were placed.

Severe myocardial ischemia was induced in pigs by clamping of the distal third of the left anterior descending coronary artery (LAD). Postmortem determinations revealed that about 15% of both ventricles were ischemic. All experiments were performed in NLA-anesthesia. Further details of the methods employed in our laboratory have been described recently in detail (5).

In open chest NLA anesthetized dogs, light, moderate or severe myocardial ischemia was induced by clamping of one, or more side branches of the LAD. The same type of ion-selective electrodes (Fig. 1) as used in the pig experiments were sewn on the epicardium of the left ventricle with atraumatic threads as close as possible to the LAD.

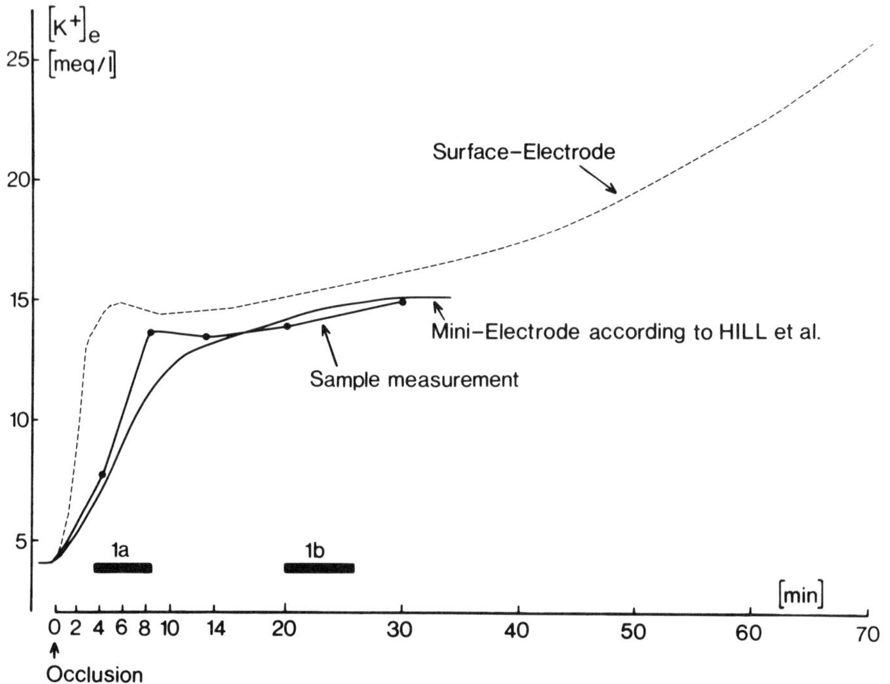


Fig. 2. Increase of extracellular $K^+$ concentration ($[K^+]$e) measured in the centre of the ischemic region of the left ventricle of pigs with: surface electrodes described in Fig. 1 (-----------) (17 pigs); with mini-electrodes according to Hill et al. (4) (——————) (3 pigs); and in samples of extracellular fluid absorbed from the epicardium with filter paper discs ( -•- ) (2 pigs). All pigs developed ventricular arrhythmias during the early phases of arrhythmia (marked as 1a and 1b) and died from ventricular fibrillation at the end of the second early phase of arrhythmia (1b). For further explanation see text

## Results and Discussion

Mean values of $[K^+]e$ measured with our surface-electrodes in the centre of the ischemic region of 17 pigs are shown in Fig. 2. $[K^+]e$ increased within 4 min up to nearly 15 meq/l. Then it decreased slightly and started to increase again after 10 min reaching 26 meq/l after 70 min. $[K^+]e$ measured with mini-electrodes according to Hill et al. (3, 4) increased also in the centre of the ischemic region but at a much lower rate of $[K^+]e$ rise, reaching 15 meq/l after about 25 min of ischemia. Using our filter paper disc method $[K^+]e$ was found to increase up to 14 meq/l after 8 min and up to 15 meq/l after 30 min of coronary artery occlusion.

A similar time course of the increase of $[K^+]e$ in the centre of the ischemic region with a phase of rapidly rising $[K^+]e$, a plateau phase and a phase of slowly rising $[K^+]e$ of pig hearts was recently described by Hill et al. (3). Although these authors (3) described also $[K^+]e$ increases up to 16 meq/l during the plateau phase of $[K^+]e$ increase in some experiments, the mean values of the rate of $[K^+]e$ rise and of $[K^+]e$ values during the plateau phase were lower than ours. Whether these differences were caused by the different types of ion-selective electrodes, by different anesthesia used in the experiments, by different size of the ischemic regions, or other factors cannot be explained at present.

In Fig. 3 results of simultaneous measurements of $[K^+]e$ and $[Na^+]e$ on anesthetized dogs with ion-selective electrodes as described in Fig. 1 are shown. Light, moderate or severe myocardial ischemia was induced by clamping of a different number of side branches of the LAD. The height of $[K^+]e$ increase was considered to reflect the degree of myocardial ischemia. During light and moderate myocardial ischemia (curves 1, and 2, Fig. 3), $[Na^+]e$ decreased transiently and returned to control values after 22 - 24 min of ischemia. During severe ischemia (curves 3, and 4, Fig. 3), $[Na^+]e$ at first transiently decreased, however, it started to increase after 2 - 5 min reaching maximal values of 165 meq/l after 10 min of ischemia.

In pig experiments $[Na^+]e$ also increased after 8 min of ischemia up to 180 meq/l then it decreased slowly by about 20 meq/l during the following 50 min of ischemia. $[Ca^{2+}]e$ increased reaching 3.9 meq/l after 30 min of ischemia. $[H^+]e$ increased with a constant rate of about 50 neq/l/min reaching nearly 3000 neq/l after 60 min of ischemia.

The time course of $[K^+]e$ increase, with a rapid $[K^+]e$ increase, a plateau phase but no further $[K^+]e$ increase observed in our dog experiments (Fig. 3) is consistent with the time course of $[K^+]e$ increase which was found also in dogs by Wiegand et al. (8). One must therefore assume that in the ischemic pig heart after the plateau phase, $[K^+]e$ slowly further increases (3, 5, 6, 9) while it does not in the dog (8, Fig. 3).

The increase of $[Na^+]e$ and $[Ca^{2+}]e$ may be explained by a shrinkage of the extracellular space (ECS) of the ischemic myocardium which is due to a $H_2O$-shift from the ECS to the intracellular space (ICS). This transmembrane $H_2O$-shift was calculated from the increase of the extracellular osmolality. The extracellular osmolality increased within 40 min from 290 up to 440 mosmol/kg. For calculation of the changes of the extra- and intracellular $H_2O$ space from the extracellular osmolality it was assumed, first that no $H_2O$ diffused from normal perfused myocardium to the ischemic area and second

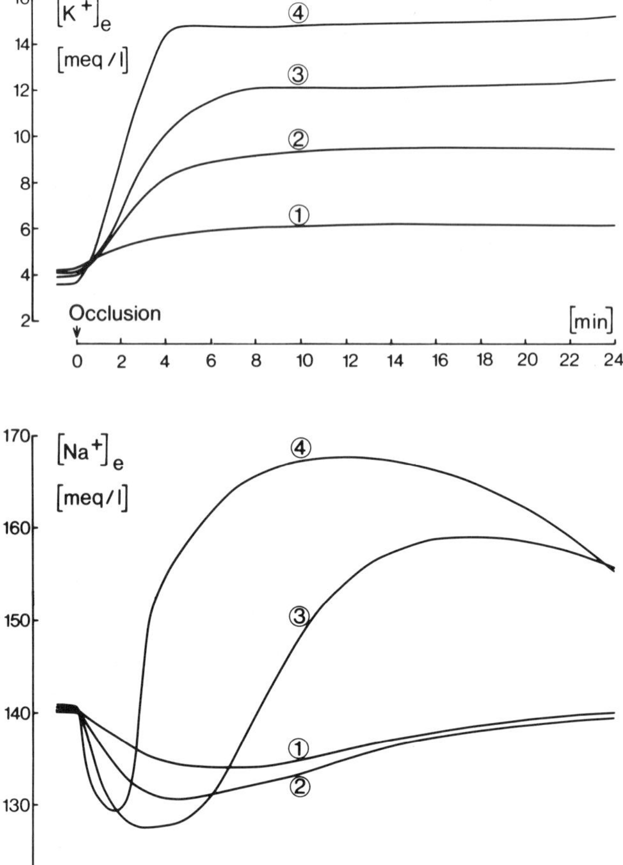



Fig. 3. Changes of extracellular $K^+$- ($[K^+]e$) and $Na^+$-concentration ($[Na^+]e$) measured with surface electrodes as described in Fig. 1 during light, moderate and severe ischemia. The height of $[K^+]e$ increase in the ischemic region was taken as measure of the degree of ischemia. 12 ischemias were induced in three different dog experiments. For further explanation see text

that the number of osmotic active particles in the extracellular space remained constant. These calculations revealed that the ECS decreased by about 32% while the ICS increased by 13% within 40 min of ischemia.

Considering the different ion-fluxes and the increase of the intracellular $H_2O$ content; the changes of the intracellular concentrations of $Na^+$ and $K^+$ ($[Na^+]i$, $[K^+]i$); and from the changes of $[Na^+]e$ and $[K^+]e$ and $[Na^+]i$ and

$[K^+]_i$, the changes of the $Na^+$- and the $K^+$-equilibrium potentials ($E_{Na^+}$, $E_{K^+}$) could be calculated. While $E_{Na^+}$ increased transiently from +60 to +74 mV and then decreased to its preocclusion value 40 min after the onset of occlusion; $E_{K^+}$ decreased from -94 to -57 mV within 4 min of ischemia.

## Summary

In open chest pigs during severe ischemia the extracellular concentrations of $Na^+$, $K^+$, $Ca^{2+}$ and $H^+$ increase. Extracellular osmolality of the ischemic area increases due to a $H_2O$-shift from the ECS to the ICS. The increase of $[Na^+]e$ must be explained by the considerable shrinkage of the ECS which is due to the $H_2O$-shift. The increase of $[Ca^{2+}]e$ is additionally caused by the decrease of pHe. The increase of the $[K^+]e$ is mainly caused by the release of $K^+$ from the ICS. The considerable changes of $[K^+]e$ and $[K^+]i$ cause a decrease of the membrane potential to a range in which slow response potentials and re-entry excitations can occur. The increase of the $[K^+]e$ therefore seems to be an important factor causing early postischemic arrhythmias.

## References

1. Ammann D, Bissig R, Cimermann Z, Fiedler U, Güggi M, Morf WE, Oehme M, Osswald H, Pretsch E, Simon W (1976) Synthetic neutral carriers for cations. In: Kessler M, Clark LC Jr, Lübbers DW, Silver IA, Simon W (eds) Ion and enzyme electrodes in biology and medicine. Urban & Schwarzenberg, München Berlin Wien, p 22
2. Gebert G, Benzing H, Strohm M (1971) Changes in the interstitial pH of dogs myocardium in response to local ischemia, hypoxia, hyper-, and hypocapnia, measured continuously by means of glass microelectrodes. Pfluegers Arch 329:72-81
3. Hill JL, Gettes LS (1980) Effect of acute coronary artery occlusion on local myocardial extracellular $K^+$ activity in swine. Circulation 61: 768-778
4. Hill JL, Gettes LS, Lynch MR, Hebert NC (1978) Flexible valinomycin electrodes for on-line determination of intravascular and myocardial $K^+$. Am J Physiol 235:H455
5. Hirche H, Franz C, Bös L, Bissig R, Lang R, Schramm M (1980) Myocardial extracellular $K^+$ and $H^+$ increase and noradrenaline release as possible cause of early arrhythmias following acute coronary artery occlusion in pigs. J Mol Cell Cardiol 12:(in press)
6. Hirche H, Bovenkamp U, Franz C, Gaetgens P, Hagemann H, Kebbel U, Schramm M, Schumacher E (1977) Changes of local myocardial contractility and extracellular $H^+$- and $K^+$- activity during ischemia. Proc Int Un Physiol Sc Vol XIII, No 951, p 324
7. Höper J, Kessler M, Simon W (1976) Measurements with ion-selective surface electrodes (pK, pNa, pCa, pH) during no-flow anoxia. In: Kessler M, Clark LC Jr, Lübbers DW, Silver IA, Simon W (eds) Ion and enzyme electrodes in biology and medicine. Urban & Schwarzenberg, München Berlin Wien, p 331
8. Wiegand V, Güggi M, Meesmann W, Kessler M, Greitschus F (1979) Extracellular potassium activity changes in the canine myocardium after acute coronary occlusion and the influence of ß-blockade. Cardiovasc Res 13: 297-302

9. Zylka V, Friedrich R, Bissig R, Hirche H, Schonert S (1980) Myocardial electrolyte shifts following coronary artery occlusion. Proc XXVIII Int Congr Physiol Sc Vol XIV, No 3757, 802

Acknowledgements

Supported by the Deutsche Forschungsgemeinschaft, SFB 68, A 7

Universität Köln, Lehrstuhl für Angewandte Physiologie, Robert-Koch-Str. 39, 5000 Köln 41, FRG

Discussion
==========

Hill:  1) I would like to point out that another possibility exists for differences in the magnitude and rate of the $K^+$ rise measured by intramuscular and surface electrodes in our laboratories. That is, the role played by lymphatic flow in the drainage of the extracellular space, particularly at the epicardial surface. A surface probe as large as 1 cm in diameter that is sewn to the surface would produce a significant force on the walls of surface lymphatic vessels, reducing or eliminating the lymphatic flow in that area and thus removing a potassium or ionic sink which would be operable under more physiological conditions. Results from our experiments using $K^+$ dlectrodes subendocardium and subepicardium at the same site in the ischemic zone indicate that subepicardial $K^+$ rises more slowly and to a lesser degree than in the subendocardium (see Fig. 2 of the following paper). The lymphatic drainage at the epicardium may be a possible explanation for this latter observation.  2) The position of the reference electrode is critical for accurate measurements with ion-selective electrodes in ischemic tissue. The reference must be very close to the ion-selective membrane in order to avoid measurement of DC electrical potentials (known to arise during myocardial ischemia) which would be misinterpreted as a change in ionic activity. This is another potential source for differences between these two methods.

Hirche:  ad 1) The weight of the trovidur-plate containing 6 ion-selective electrodes is about 1 g and this small force is exerted on an area of the epicardium of about 0.8 $cm^2$. The fixation of the trovidur-plate on the epicardium with three atraumatic threads is somewhat elastic. When a bolus of 2.5 meq of KCl is injected into the right atrium before clamping of the LAD after the peak $K^+$ concentration is exceeded $[K^+]_e$ decreases as rapidly as the $K^+$ concentration measured simultaneously with catheter electrodes in the coronary sinus. Therefore, we don't have any evidence that local lymphatic drainage is impeded by our surface electrodes.  ad 2) We have measured DC-potentials up to 4 mV after 5 min of occlusion between the calomel electrode (reference electrode for all ion-selective electrodes) placed between thoracid muscles and surface reference electrodes sewn to the epicardium of the ischemic region. This means that $[K^+]_e$ can be measured by up to 2.0 meq/l too high at a $[K^+]_e$ of 15 meq/l. The slower increase of $[K^+]_e$ during the first min after the onset of occlusion which we have also seen using $K^+$ selective mini-electrodes according to Hill et al. (see Fig. 2) possibly might be due to the local damage of tissue at the tip of these electrodes which cause an enlargement of the artificial extracellular space. This would increase the diffusion distance of $K^+$ from intact cells to the tip of the electrode during ischemia.

# The Use of K$^+$ Sensitive Electrodes to Gain an Understanding of Myocardial Ischemia

L.S. GETTES, J.L. HILL, E. NORFLETE, G.F. LOPEZ

## Introduction

Since the early work of Harris and co-workers (4), it has been known that the acute interruption of coronary blood flow results in the appearance of potassium in the vein draining the ischemic area and in the coronary sinus. This early work and the subsequent studies which evolved from it established the relationship between ischemia and an increase in extracellular potassium and suggested that the accumulation of potassium was an important factor in the development of the electrical and mechanical disorders which characterize myocardial ischemia. It is readily apparant, however, that the appearance of potassium in  the coronary sinus or coronary vein and the magnitude of the change in coronary venous potassium are influenced by several factors, including the following: 1) the mass of the ischemic myocardium, 2) the magnitude of the change in extracellular K$^+$ in the ischemic myocardium, and 3) the volume of blood from nonischemic areas which must be present for any flow to occur in the vein draining the ischemic area.

The development of potassium sensitive electrodes which can be inserted into the myocardium of beating hearts provides the opportunity to characterize the ischemic process by observing the changes in extracellular potassium within the various regions of the ischemic zone. Such an electrode would permit the derivation of the following information: 1) The time course and magnitude of the potassium change following acute occlusion and release, 2) the determination of inhomogeneities within the ischemic area, 3) the correlation of the changes in extracellular potassium to changes in electrical and mechanical properties, 4) the determination of the role of potassium changes in the production of the associated electrical and mechanical changes, 5) the relationship of changes in potassium to changes in other ions and to changes in blood flow, and 6) the effects of physiological and pharmacological interventions on the above properties.

We have studied many of the correlations mentioned above using a previously described PVC-valinomycin membrane system (6, 7). The electrodes were inserted at various locations within the ischemic zone. These included the midmyocardial center of the ischemic zone, the subepicardial and subendocardial layers of the ischemic zone, the margin of the ischemic zone, as defined by the visible cyanotic border and the nonischemic myocardium. Unipolar electrograms were recorded simultaneously with the K$^+$ signal from the reference barrel of the K$^+$ electrode versus a central terminal. In some experiments, an additional reference electrode was positioned within 1 mm of the K$^+$ electrode tip creating a triple barrel electrode. The two reference barrels were then used to record bipolar electrograms. Midmyocardial and epicardial temperatures within the anticipated ischemic zone were also monitored using Yellow Springs Instruments Model 511 flexible probes for intramyocardial recordings and a ≠≠ 42112-4 surface probe for surface temperature recordings. In some experiments, coronary blood flow was determined using radiolabeled microsphere techniques. The microspheres were 8 - 10 /um in diameter and were injected into the left atrium. In general, the animals were anesthetized using pentobarbital, 30 mg/kg or alpha chloralose. In some experiments, halothane inhalation anesthesia was utilized.


Results

Figure 1 demonstrates the time course of the potassium change as recorded by two electrodes inserted in the midmyocardium of the center of the ischemic zone, a third at the inner margin of the ischemic zone, and a forth electrode in the non-ischemic zone. The ordinate shows both the measured potassium activity and the calculated potassium concentration. The figure demonstrates that the potassium rises rapidly after the ligation of the left anterior descending coronary artery reaching a plateau level after 8 - 10 minutes. The plateau phase lasts 10 - 15 minutes and is followed by a phase of slowly rising $K^+$ which reaches 21 mM after 60 minutes. At the margin, the potassium rise begins somewhat later than the rise in the center. It rises more slowly and to lower levels than in the center of the ischemic zone reaching 13 mM after 60 minutes of occlusion. Experiments of this type in which the ligation was maintained for more than three hours demonstrated a peak potassium in the center of the ischemic zone of over 40 mM. Figure 1 also shows that major inhomogeneities exist not only between the center and the margin of the ischemic zone, but also between electrodes placed in the midmyocardium of the center of the ischemic zone. Major inhomogeneities also exist between the subendocardial and subepicardial regions in the center of the ischemic zone. This is shown in Figure 2 which also demonstrates significant inhomogeneities after the occlusion is released and perfusion is restored. Thus, the use of the $K^+$ electrode clearly indicates that major inhomogeneities of the ischemic process exist throughout the ischemic zone and not merely at the margin as had been previously determined using various other techniques such as radioactive microspheres (1, 10, 12, 14) and various anatomic (10) and metabolic (9, 11, 5) parameters.

The use of the $K^+$ electrode also permits the correlation between the changes in extracellular potassium and ventricular activation which occur following coronary artery occlusion. Figure 3 demonstrates that the inhomogeneities of the changes in extracellular potassium are paralleled by inhomogeneities in the electrical activity as recorded by either unipolar or bipolar electrodes. This figure shows bipolar recordings in panel A and unipolar recordings in panel B from two different experiments. In panel A, the margin and central zones are compared. In panel B, the signals recorded from electrodes placed in the center between closely placed electrodes in the center of the ischemic zone are compared.

Our previously described work confirms the observations of others (3, 13) indicating that changes in potassium cannot be the sole cause of the activation changes induced by acute coronary ligation. Nonetheless, the fact that the activation changes parallel the $K^+$ rise permit the utilization of this technique as a qualitative and perhaps a semi-quantitative guide to the associated activation abnormalities.

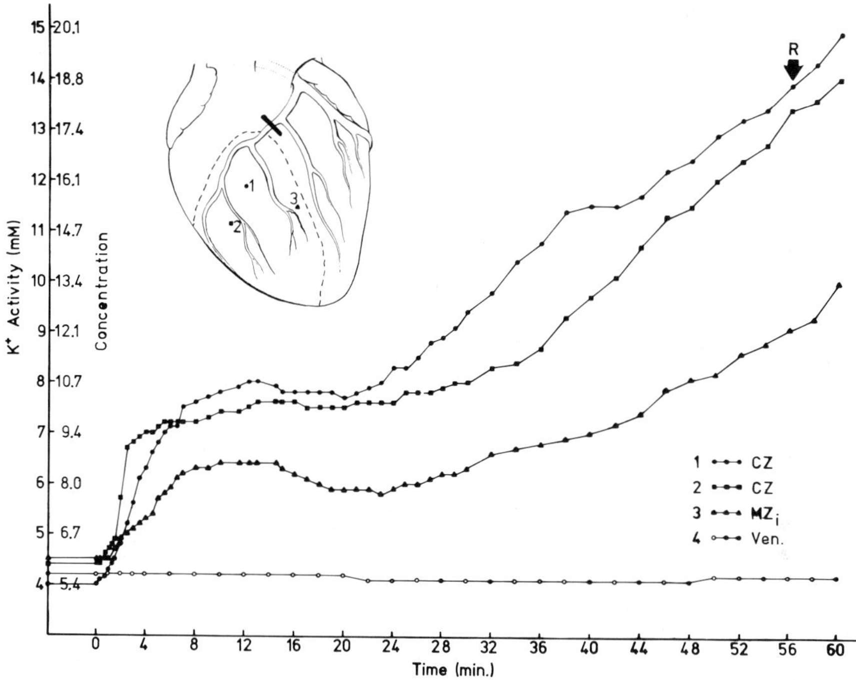

Fig. 1 . Effect of occlusion of the left anterior descending coronary artery in  the pig on the extracellular $K^+$ change recorded from midmyocardial electrodes positioned as shown in the inset. CZ = the center of the
ischemic zone, MZ = inside margin of the ischemic zone, i.e. less than 5 mm
from the visible margin of the ischemic zone. NZ = nonischemic zone. The
arrow at 56 minutes marks the release of the occlusion. Discussion in text.
Reprinted from Hill and Gettes (7)


The relationship of the changes in potassium to other indices of ischemia
indicate that although slight differences may be observed, the potassium
electrode is an excellent marker of the ischemic process. Figures 1 and 2
illustrate that the changes in potassium are similar to the changes in $pCO_2$
recorded by Case et al. (2). However, the changes in $pCO_2$ differ from the
changes recorded by the potassium electrode in the absence of a plateau
phase. This absence of a plateau phase is noteworthy since in our experiments, the end of the plateau phase identified what may be the end of the
period of reversible ischemia. Thus, the correlation of the data from the
potassium electrode to the data obtained by other means suggests that the
$K^+$ electrode is a sensitive and perhaps a better index of ischemia than
other available technologies. It has thus been possible to utilize the

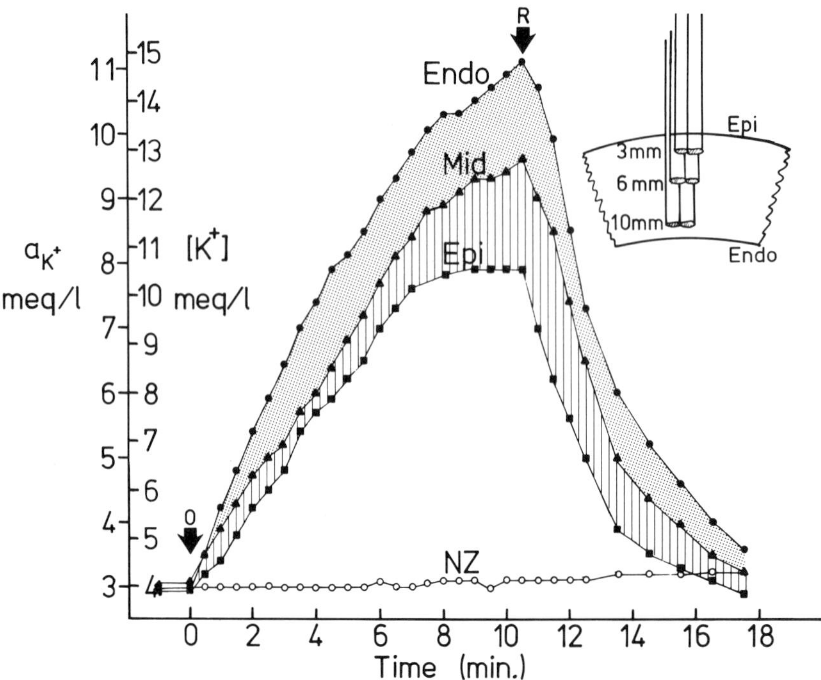


Fig. 2. Effect of occlusion and release of left anterior descending coronary artery in the pig on extracellular $K^+$ change recorded from subepicardial, midmyocardial, and subendocardial $K^+$ electrodes. Note that the difference between $K^+$ levels at subendocardium and subepicardium approaches 5 mM. Discussion in text

potassium electrode to determine the effects of physiological and pharmacological interventions which might modify the ischemic process.

Figure 4 illustrates an experiment in which we studied the effects of changes of halothane concentration on midmyocardial ischemia induced by an 85% reduction in flow through the LAD coronary artery. In this experiment, a femoral-coronary artery shunt was installed to allow a gradual decrease in coronary flow so as to establish the ischemia threshold. The flow rate was then maintained at this level and halothane concentration was changed. This resulted in a decrease in arterial pressure and in dp/dt of ventricular pressure. The extracellular potassium falls in the jeopardized areas, but extracellular potassium rises in the formally nonischemic area. This result suggests that a decrease in cardiac output and coronary flow was associated with the decrease in left ventricular systolic pressure and

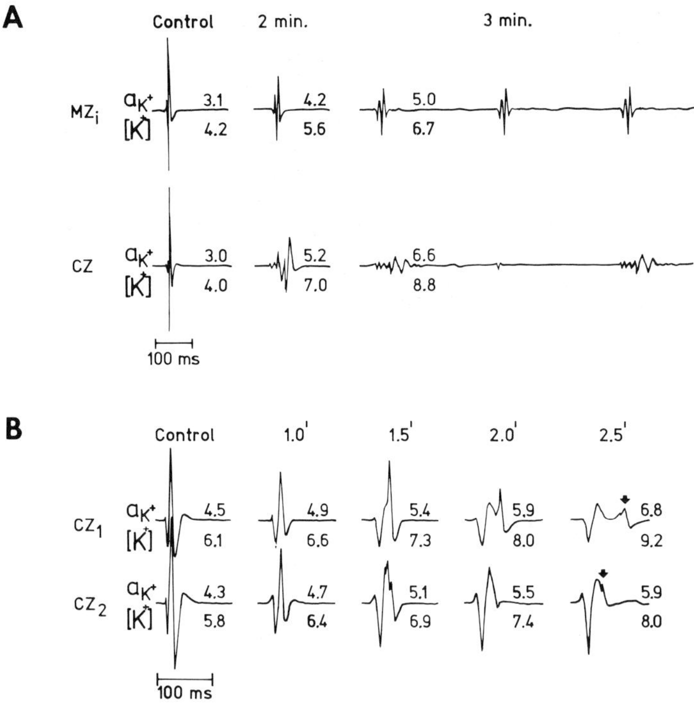

Fig. 3.  Heterogenous changes in ventricular activation recorded from the
tip of midmyocardial $K^+$ electrodes after left anterior descending coronary
artery occlusion  in the pig. Corresponding values of extracellular $K^+$ activity ($ak^+$) and calculated equivalent $K^+$ concentration are shown. In A,
bipolar electrograms recorded before, two and three minutes following occlusion are shown. Note that at 2 minutes following occlusion there is a
significant delay in activation at CZ but only a slight delay at MZ. By
three minutes, the activation delay at MZ is more pronounced while at CZ, a
2:1 block in activation has occurred, and the spikes of the conducted impulses show marked delay. In B, unipolar electrograms recorded from two
electrodes positioned 1 cm apart in the center of the ischemic zone are
shown. $K^+$ activity rises faster and to a higher level in the local activation spike. Reprinted from Hill and Gettes (7)


dp/dt producing ischemia in the normal zone whereas, in the jeopardized
area where flow is maintained at 15% of the control value, the decrease in
cardiac work resulted in a lessening of myocardial ischemia. Other work in
our laboratory has established that changes in heart rate (7), beta adren-
ergic stimulation and blockade (Lopez, S.F., Hill, J.L. and Gettes, L.S.,
unpublished observation) are incapable of altering the rate of rise
or the magnitude of $K^+$ in the center of the ischemic zone.

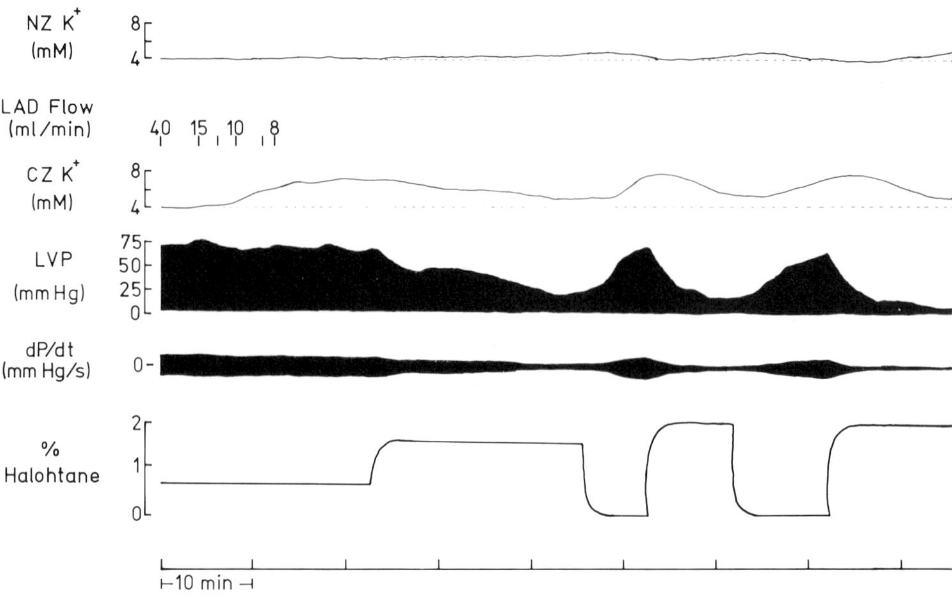



Fig. 4. Results of an experiment in which the effect of changes in Halothane concentration on subtotal myocardial ischemia are studied using midmyocardial $K^+$ electrode positioned in the ischemic and non-ischemic zones. LAD flow is via a femoral artery - LAD shunt and is regulated by a roller pump. NZ = non-ischemic zone $K^+$. CZ = ischemic zone $K^+$. LVP = left ventricular pressure. Note that the first rise in $K^+$ in the ischemic area occurs when coronary flow is reduced from 40 to 12 ml per minute. As flow is further reduced to 8 ml per minute $K^+$ rises more and is constant at this level. At 22 minutes, halothane concentration is increased from 0.75 to 1.5%. This results in a decrease in left ventricular pressure and in dp/dt. This is associated with a decrease in midmyocardial potassium in the nonischemic area. At 36 minutes, the halothane is decreased from 1.5 to 0%. This results in a restoration of the left ventricular pressure and dp/dt, a decrease in the nonischemic zone $K^+$ to the control levels and an increase in the ischemic zone $K^+$ implying increasing ischemia. The sequence is then repeated as halothane is raised and lowered sequentially. Discussion in text

## Discussion

The results of these experiments indicate that the valinomycin-PVC matrix
membrane electrode is a most useful technique to gain an improved understanding of myocardial ischemia. Our results and those of Wiegand et al.
(14) and Hirche et al. (8) using surface electrodes invite many provocative
questions. For instance: 1) What is the precise cause of the initial and
subsequent $K^+$ rise and of the plateau phase? 2) Why do inhomogeneities exist even in the center of the ischemic zone? 3) What are the roles of the
magnitude, rate and inhomogeneities of the $K^+$ rise on the associated electrical and mechanical changes? 4) Do the inhomogeneities in K following
restoration of coronary flow contribute to reperfusion arrhythmias? And 5)
Can the $K^+$ changes be modified?

The differences in the $K^+$ values reported by us and Hirche in the preced-
ing paper (5) suggest that one or both of the techniques may not be per-
fectly accurate. In our works the effects of a possible artificial space
created by the electrodes might result in the measured $K^+$ change being
slightly low. However, the in vivo calibration described and illustrated in
our earlier reports (6, 7) indicate that this error must be quite small. In
the work of Hirche et al. (5), possible errors introduced by the use of
the larger distant reference electrode and possible changes occurring under
the surface of the ion-specific electrode must be considered. None the
less, both studies show that $K^+$ rises to levels which are higher than
previously reported.

Our results indicate that the $K^+$ changes in the center and border correlate to results obtained using other techniques (1, 5, 9, 10, 11, 12).
However, our results reveal greater inhomogeneities in $K^+$ and, by implication, in the ischemic process than can be recorded by these other averaging
techniques. These inhomogeneities are consistent with inhomogeneities in
electrophysiologic markers shown by others (3) and confirmed in our studies. Our previously published results (2) are also consistent with the
conclusion that the $K^+$ changes, of themselves, cannot explain the associated electrical abnormalities. None the less, our study indicates that the
$K^+$ changes are paralleled by activation changes and may thus be useful as
a predictor of these changes.

The use of the $K^+$ electrode appears to be a most useful means of assessing
the results of physiologic, pharmacologic and surgical interventions. As
such, the electrodes may provide a means of evaluating the effects of
interventions designed to lessen the extent of infarction or to protect the
ischemic myocardium. Moreover, the extension of this technology to the
clinical setting may provide the means of studying myocardial ischemia in
man.

## References


1. Becker LC, Fortuin NJ, Pitt B (1971) Effect of ischemia and antianginal
   drugs on the distribution of radioactive microspheres in the canine
   left ventricle. Circ Res: 263
2. Case RB, Felix A, Castellana FS (1979) Rate of the rise of myocardial
   $pCO_2$ during early myocardial ischemia. Circ Res 45: 324

178


3. Downar E, Janse MJ, Durrer D (1977) The effect of acute coronary artery occlusion on subepicardial transmembrane potentials in the intact porcine heart. Circulation 56: 21
4. Harris AS, Bisteni A, Russel RA, Brigham JC, Firestone JE (1954) Excitatory factors in ventricular tachycardia resulting from myocardial ischemia: potassium a major excitant. Science 119: 200
5. Hearse DJ, Opie LH, Katzeff IE, Lubbe WF, VanDer Werff TJ, Baulle G (1977) Characterization of the "border zone" in acute regional ischemia in the dog. Am J Cardiol 40: 716
6. Hill JL, Gettes LS, Lynch MR, Hebert NC (1978) Flexible valinomycin electrodes for on-line determination of intravascular and myocardial $K^+$. Am J Physiol 235: H455
7. Hill JL, Gettes LS (1980) Effect of acute coronary artery occlusion on local mayocardial extracellular $K^+$ activity in swines. Circulation 61: 768
8. Hirche et al (1980) Measurements of myocardial extracellular $Na^+$, $K^+$, $Ca^{2+}$ and $H^+$ using ion-selective electrodes during ischemia. Symposium procedings
9. Janse MJ, Cinca J, Morena H, Fiolet JWT, Kleber AG, deVries GP, Becker AE, Durrer D (1979) The "border zone" in myocardial ischemia: an electrophysiological, metabolic, and histochemical correlation in the pig heart. Circ Res 44: 576
10. Jennings RB, Ganote CE, Reimer KA (1975) Ischemic tissue injury. Am J Pathol 81: 179
10. Jennings RB, Ganote CE, Reimer KA (1975) Ischemic tissue injury. Am J Pathol 81: 179
11. Karlsson J, Templeton GH, Willerson JT (1973) Relationship between epicardial S-T segment changes and myocardial metabolism during acute coronary insufficiency. Circ Res 32: 725
12. Marcus ML, Kerber RE, Ehrhardt J, Abboud RM (1975) Three dimensional geometry of actuely ischemic myocardium. Circulation 52: 254
13. Morena H, Ganse MJ, Fiolet ZWT, Krieger WJG, Crijns H, Durrer D (1980) Comparison of the effects of regional ischemia, hypoxia, hypercapnia and acidosis on intracellular and extracellular potentials and metabolism in the isolated porcins heart. Circ Res 46: 634
14. Wiegand V, Güggi M, Meesmann W, Kessler M, Greitschus F (1979) Extracellular potassium activity changes in the canine myocardium after coronary occlusion and the influence of beta blockade. Cardiovasc Res 13: 297



This work supported by NHLBI grant HL23624.



Department of Medicine, Division of Cardiology, University of North Carolina at Chapel Hill, Chapel Hill, North Carolina, USA 27514

# Intracellular Potassium Activity in Normal and Hypoxic Guinea Pig Papillary Muscle

J.L. WALKER, W.G. WIER

It has been generally accepted by cardiac electrophysiologists that the diastolic (resting) membrane potential, $E_D$, of myocardial cells is determined, primarily, by the potassium equilibrium potential, $E_K$ (2). This has been shown to be true for myocardial cells in which intracellular potassium activity, $a_i(K)$, has been measured with potassium selective microelectrodes (4, 8). McDonald and MacLeod, however, published experimental results (5), which they later expanded (6), from which they concluded that under some conditions $E_D$ is not primarily determined by $E_K$. They superfused guinea pig papillary muscles under nitrogen and found there was a large loss of intracellular potassium but $E_D$ did not increase as much as $E_K$, i.e. $E_D$ became negative with respect to $E_K$. They concluded that during hypoxia, $E_D$ was maintained negative with respect to $E_K$ by electrogenic sodium pumping. Bosteels et al. (2) did similar experiments with embryonic chick heart and found, to the contrary, that $E_D$ always remained positive with respect to $E_K$. We have repeated the experiments with guinea pig papillary muscles using potassium selective electrodes, instead of whole tissue analysis, to measure intracellular potassium.

Methods

Guinea pigs were killed by cervical dislocation, the heart rapidly removed and rinsed with Kreb's solution. A papillary muscle was removed from the right ventricle and fixed to the bottom of a tissue bath with a pin through each end. The muscle was stimulated at a rate of 1 Hz while being superfused with Kreb's solution at $37^{O}$C. One hour was allowed for equilibration of the muscle in the bath, after which time control measurements of $E_D$ and $a_i(K)$ were made. Following the control measurements, during which time the Kreb's solution was equilibrated with 95% $O_2$-5% $CO_2$, the muscle was made hypoxic by superfusing it with Kreb's solution equilibrated with 95% $N_2$-5% $CO_2$. $E_D$ and $a_i(K)$ were measured again after two hours of hypoxia, then oxygen was returned and the measurements again repeated after fifteen minutes of recovery.

$E_D$ measurements were made with 3 M KCl filled micropipettes with resistances in the range of 12 - 15 M$\Omega$. Potassium selective microelectrodes were made by treating the pipettes with Siliclad (Clay-Adams) and then filling the tip with potassium liquid ion-exchanger (Corning code 477317) (7). The potassium electrodes were calibrated with pure KCl solutions and the selectivity with respect to sodium was determined by referring the electrode potential measured in Kreb's solution, to the KCl calibration curve. Calibration was performed before and after each group of measurements. Five measurements were made with each of the two kinds of electrodes and the means of the measurements were used to calculate $a_i(K)$ from which $E_K$ could in turn be calculated. Standard deviations were calculated by a propagation of error method (3). $a_i(K)$ was calculated using equation 1.

$$a_i(K) = a'_o(K) \exp\left[\frac{(\Delta E - E_D)F}{RT}\right] \tag{1}$$

$a_i(K)$ (mM) is the intracellular potassium activity, $a'_o(K)$ (mM) is the apparent extracellular potassium activity as defined by equation 2; $\Delta E$ (mv) is the potential difference between the outside and inside the cell during diastole, measured with the potassium electrode; $E_D$ (mv) is the diastolic membrane potential measured with the KCl filled pipette ($E_M$ electrode); and R, T and F have their usual meanings.

$$a'_o(K) = a_o(K) + k_{K,Na}a_o(Na) \tag{2}$$

$a_o(K)$ (mM) is the potassium activity in the Kreb's solution, and $k_{K,Na}$ (dimensionless) is the selectivity coefficient for potassium and sodium. $k_{K,Na}$ was in the range of 0.01 - 0.02 which was sufficiently small to justify ignoring the contribution of $a_i(Na)$ to the potential of the potassium electrode. Electrode potentials were either read directly from the oscilloscope or from the film record made by photographing the oscilloscope. Intracellular measurements were made only in superficial cells of the muscle.

## Results

Table 1 shows the results of the experiments in which $a_i(K)$ was measured in normal and hypoxic guinea pig papillary muscles. The values presented in the table are means and standard deviations. In eight control muscles, superfused with oxygenated Kreb's solution, the mean diastolic membrane potential, $E_D$, was -96.6 mv and the mean $a_i(K)$ was 99.7 mM. The potassium activity in the Kreb's solution was 2.2 mM and the calculated mean $E_K$ was -100.7 mv. Five of the muscles were superfused with nitrogenated Kreb's solution, and in these $E_D$ increased to -90 mv while $a_i(K)$ decreased to 86 mM and $E_K$ increased to -96.9 mv. In all cases $E_K$ remained negative with respect to $E_D$. In four of the five muscles, intracellular measurements were obtained during the first fifteen minutes of recovery, i.e. after oxygenated Kreb's was readmitted. During recovery $E_D$ and $a_i(K)$, and therefore also $E_K$, returned to near their control values.

Table 1. Membrane potential, intracellular potassium activity and potassium equilibrium potential in guinea pig papillary muscle

| | n | $E_D$ (mv) | $a_i(K)$ (mM) | $E_K$ (mv) |
|---|---|---|---|---|
| CONTROL | 8 | -96.1 $\pm$ 2.3 | 99.7 $\pm$ 11 | -100.7 $\pm$ 3.4 |
| HYPOXIC (2 HRS) | 5 | -90.0 $\pm$ 3.4 | 86.2 $\pm$ 7 | -96.1 $\pm$ 2.2 |
| RECOVERY (15 MIN) | 4 | -98.0 $\pm$ 3.2 | 107.8 $\pm$ 24 | -102.5 $\pm$ 6.2 |

The changes from the control to the hypoxic values of $E_D$ and $a_i(K)$ were significant at the levels of 0.005 and 0.025 respectively. The values of $E_D$ and $a_i(K)$ during recovery were not significantly different from the control values.

## Discussion

For comparison with our results, those of McDonald and MacLeod (6) are summarized in Table 2. Whereas their results show a greater increase in $E_K$ than in $E_D$, so that $E_K$ becomes positive with respect to $E_D$, our results show that $E_D$ always remains positive with respect to $E_K$.

Table 2.  Membrane potential, intracellular potassium concentration and potassium equilibrium potential in guinea pig papillary muscle. Summarized from McDonald and MacLeod (6)

|  | $E_D$ (mv) | $C_i$ (K) | $E_K$ (mv) |
|---|---|---|---|
| CONTROL | -81 | 106 | -83 |
| ANOXIC (8 HRS) | -77 | 27 | -47 |

We believe that the difference between the two sets of results can be explained by the difference in the methods used for measuring intracellular potassium. McDonald and MacLeod (5, 6) used whole tissue chemical analysis while we used potassium selective microelectrodes. When the preparations are made hypoxic, they can maintain near normal intracellular potassium if the glucose concentration in the Kreb's solution is increased from the normal 5 mM to 50 mM. However, when the glucose concentration is 5 mM, there is a substantial loss of intracellular potassium during hypoxia (6). It seems likely to us that in the latter case, the deeper cells in the muscle receive less glucose than the superficial cells. Therefore, their metabolism is more depressed and they lose more potassium than the superficial cells. Whole tissue analysis cannot distinguish between these different cell populations but instead, averages the potassium loss over the whole muscle. Membrane potential measurements, on the other hand, would be made in the superficial cells which, being the least depressed would be the least depolarized.

In our experiments using potassium microelectrodes, $E_D$ and $a_i$(K) were measured in the same superficial cell population. We therefore correlate the degree of depolarization and loss of potassium in the same relatively homogeneous population of cells rather than making $E_D$ measurements in a limited superficial cell population while measuring potassium loss in the whole muscle. Our conclusion, which is in agreement with that of Bosteels et al. (1), is that the diastolic membrane potential of myocardial cells, even under hypoxic conditions, is primarily a function of the potassium equilibrium potential and $E_D$ is always positive with respect of $E_K$.

## References


1. Bosteels S, Vleugels A, Carmeliet E (1975) Intracellular Na and K concentrations and membrane potential in chick embryonic heart under hypoxic conditions. Arch Intern Physiol Biochim 83:148-150
2. Coraboeuf E (1978) Ionic basis of electrical activity in cardiac tissues. Am J Physiol 234:H101-H116
3. Deming EW (1964) Statistical Adjustment of Date. Chapt. III. The propagation of error. Dover, New York

4. Fozzard HA, Lee CO (1975) Influence of changes in external potassium and chloride ion on membrane potential and intracellular potassium ion activity in rabbit ventricular muscle. J Physiol 256:663-689
5. McDonald TF, MacLeod DP (1971) Maintenance of resting potential in anoxic guinea pig ventricular muscle: Electrogenic sodium pumping. Science 172:570-572
6. McDonald TF, MacLeod DP (1973) Metabolism and the electrical activity of anoxic ventricular muscle. J Physiol 229:559-582
7. Walker JL (1971) Ion specific liquid ion exchanger microelectrodes. Anal Chem 43:89A-93A
8. Walker JL, Ladle RO (1973) Frog heart intracellular potassium activities measured with potassium microelectrodes. Am J Physiol 225: 263-267



Acknowledgements

This work was supported by National Institute of Health grants HL 18053 and NS 07938


University of Utah College of Medicine, 410 Chipeta Way, Research Park, Salt Lake City, Utah 84108/USA

# Intracellular K$^+$ Activity of Cardiac Purkinje Fibers During Temperature Change

S.S. SHEU, M. KORTH*, H.A. FOZZARD

The Na-K pump works against the electrochemical gradients of K$^+$ and Na$^+$ in order to maintain the constancy of intracellular K$^+$ and Na$^+$ activities ($a_K^i$ and $a_{Na}^i$). In addition, it has become apparent that this active transport mechanism is electrogenic and that under some conditions it can contribute directly to the membrane potential ($V_m$) (1.5). Consequently, it is important to determine the relationship between $V_m$ and the K$^+$ and Na$^+$ equilibrium potentials ($V_K$ and $V_{Na}$) when the Na-K pump is inhibited or stimulated. Using Na$^+$-selective microelectrodes, Deitmer and Ellis (2) have measured $a_{Na}^i$ during inhibition and reactivation of the Na-K pump in cardiac Purkinje fibres. The only study of $V_K$ changes in the cold, employed flame photometry for measurement of intracellular K$^+$ concentration ([K]$_i$) (11). Such an estimate of $V_K$ might not be able to account for the transmembrane potential, because of the complexity of intracellular ion distribution (9).

In the present study, K$^+$-selective microelectrodes were used to measure $a_K^i$ of cardiac Purkinje fibres when the Na-K pump was inhibited by cooling to 4$^o$C and reactivated by rewarming the fibres to 37$^o$C. This temperature change will reversibly alter the activity of the Na-K pump (11, 7), although it may also produce other effects.


## Methods

Free running Purkinje fibres, obtained from hearts of freshly slaughtered sheep, were utilized in this study. Some fibres, used in prolonged cooling experiments, were stored in oxygenated Tyrode solution containing 1.35 mM K$^+$ ([K]$_o$) at 4$^o$C for 24 hours. These fibres were then pinned in the tissue chamber during exposure to the same solution at 4$^o$C. Other fibres, which were subjected to 4 hours cooling and then rewarming, were pinned in the tissue chamber immediately after dissection. The flow rate of perfusion solutions was 10 ml/min, so that less than 10 s were required to clear the previous solution. The Tyrode solution contained (mM): 5.4 K$^+$; 161.4 Na$^+$; 2.7 Ca$^{2+}$; 1.05 Mg$^{2+}$; 149.9 Cl$^-$; 22.0 HCO$_3^-$; 2.4 H$_2$PO$_4^-$ and 11.0 glucose. Changes of K$^+$ were made by equimolar substitution for Na$^+$. All solutions used had a pH of 7.2 - 7.4 when bubbled with 95% O$_2$ and 5% CO$_2$.

Conventional 3M KCl-filled glass microelectrodes that had resistances of 10 - 30 M$\Omega$ were used to measure $V_m$. Similar microelectrodes were used for the construction of K$^+$-selective liquid ion exchanger (Corning Medical, No. 477317) microelectrodes with a method introduced by Walker (14). The electrodes were calibrated at 4$^o$C and 37$^o$C with pure solutions of KCl. The $a_K^i$ was calculated by the following equation (3):

$$E_K^i - V_m = E_O^K + S_K \log a_K^i$$

were $E_K^i$ is the transmembrane potential of the K$^+$-electrode. $E_O^K$ is the standard potential and $S_K$ is the slope calibrated with pure KCl solutions.

<u>Results</u>

Activity $a_K^i$ After 24 Hours Cooling

Fig. 1 shows the impalement of a conventional microelectrode in a fiber cooled to 4°C for more than 24 hours.

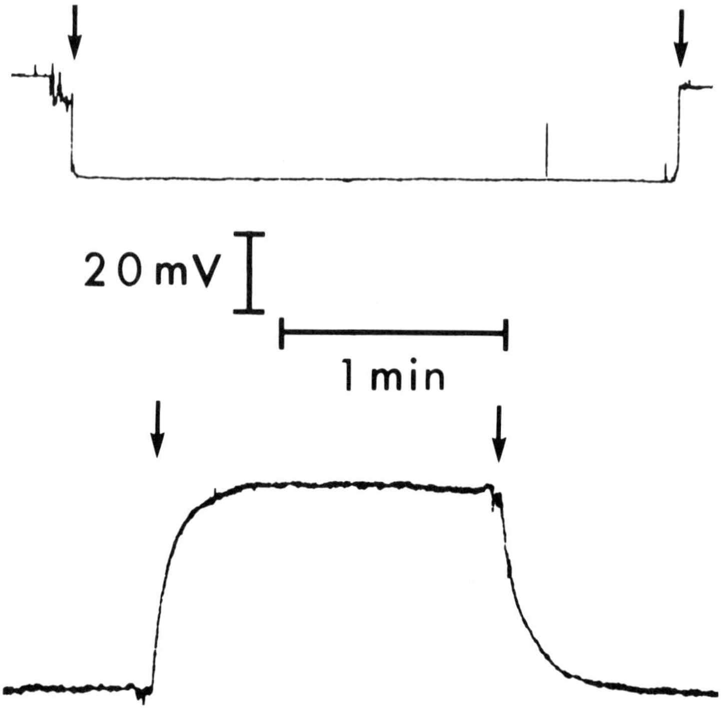


Fig. 1. Voltage recording from a conventional microelectrode (top trace) and a $K^+$-selective microelectrode (bottom trace) in a fiber kept at 4°C for more than 24 hours. The first arrow on each trace indicates the impalement and the second arrow indicates the withdrawal of the microelectrode

In 5 fibres, $V_m$ ranged from -10 mV to -25 mV, with an average value of -15.4 ± 2.8 mV (Mean ± SEM). Fig. 1 also shows the impalement of the $K^+$-electrode. The voltage difference upon impalement was a positive potential, indicating that $V_K$ was significantly more negative than $V_m$. The average $a_K^i$ in the same 5 fibres was 45.9 ± 10.1 mM. Using this number and $a_K^o$ of 0.98 mM, $V_K$ was calculated to be -89.1 ± 6.1 mV. Note that this is similar to $V_K$ in normal Tyrode solution at 37°C.

Time Course of Change in $a_K^i$ during Cooling and Rewarming

In order to study the time course of loss and recovery of $a_K^i$ during inhibition and reactivation of the Na-K pump, experiments of the sort illustrated in Fig. 2 were undertaken. $V_m$ and $a_K^i$ were first measured in Tyrode solution containing 5.4 mM $[K]_o$ at 37°C.

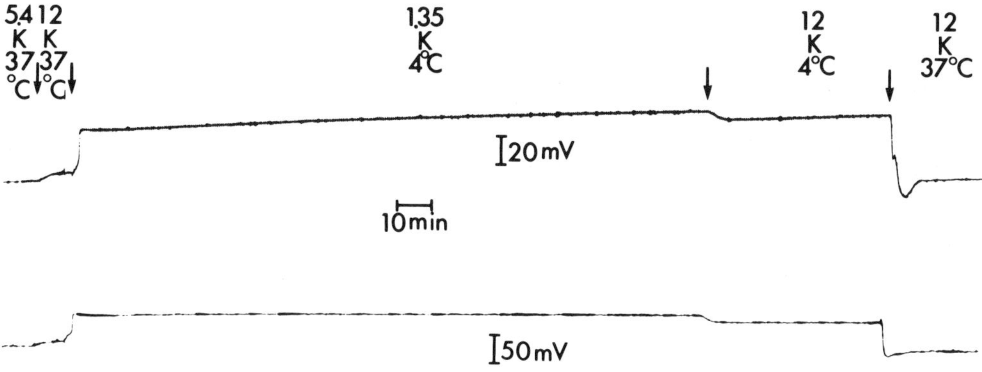


Fig. 2. Simultaneous recordings of a conventional (top trace) and a $K^+$-selective (bottom trace) microelectrode in a fibre during temperature and solution changes. For a more detailed description, see text

While the impalements were maintained, the perfusion was switched to various solutions according to the following sequences: 1.) 12 mM $[K]_o$ at 37°C for 15 minutes, 2.) 1.35 mM $[K]_o$ at 4°C for 3 hours, 3.) 12 mM $[K]_o$ at 4°C for 1 hour, and finally turned back to 12 mM $[K]_o$ at 37°C. Reasons for choosing 12 mM $[K]_o$ as the rewarming solution were that the chance of seeing hyperpolarization beyond the original $V_m$ would be higher and that the effect of $K^+$ accumulation or depletion in unstirred extracellular layers would be smaller. The results of $V_m$, $V_K$ and $a_K^i$ in 3 fibres are shown in Fig. 3.

During the initial control period of exposure to normal Tyrode solution at 37°C, $V_m$ = -79.9 $\pm$ 4.0 mV, $a_K^i$ = 106.0 $\pm$ 12.2 mM and $V_K$ = -87.8 $\pm$ 3.0 mV. Upon increase of $[K]_o$ to 12 mM, $V_m$ = -64.9 $\pm$ 1.1 mV, $a_K^i$ = 109.0 $\pm$ 11.4 mM, and $V_K$ = -67.3 $\pm$ 2.7 mV. When $[K]_o$ was switched to 1.35 mM at 4°C, there was a quick depolarization and rapid decrease of $a_K^i$ in the first 10 minutes. Afterwards, the depolarization of $V_m$ and the decrease of $a_K^i$ slowed down and appeared to reach a stable level, giving a $V_m$-$V_K$ of 80.2 mV. In this period, the average rate of depolarization for $V_m$ was 4.1 mV/h, and the rate of decrease of $a_K^i$ was 6.2 mM/h. Upon switch to 12 mM $[K]_o$ at 4°C, there was a 5 mV hyperpolarization; nevertheless, the difference between $V_m$ and $V_K$ was still more than 20 mV. During exposure to 12 mM $[K]_o$ at 37°C, $V_m$ hyperpolarized in 5 - 7 minutes to 10 mV more negative than the control value, and then returned to its original value during the next 10 - 12 minutes. In the same time, $a_K^i$ increased from 60.8 $\pm$ 13.4 mM to 120.8 $\pm$ 9.5 mM. If we assumed a surface to volume ratio of 4000 $cm^{-1}$ for the sheep cardiac Purkinje fibre (10), the potassium net influx was calculated to be 25 $pmol/cm^2 \cdot s$.

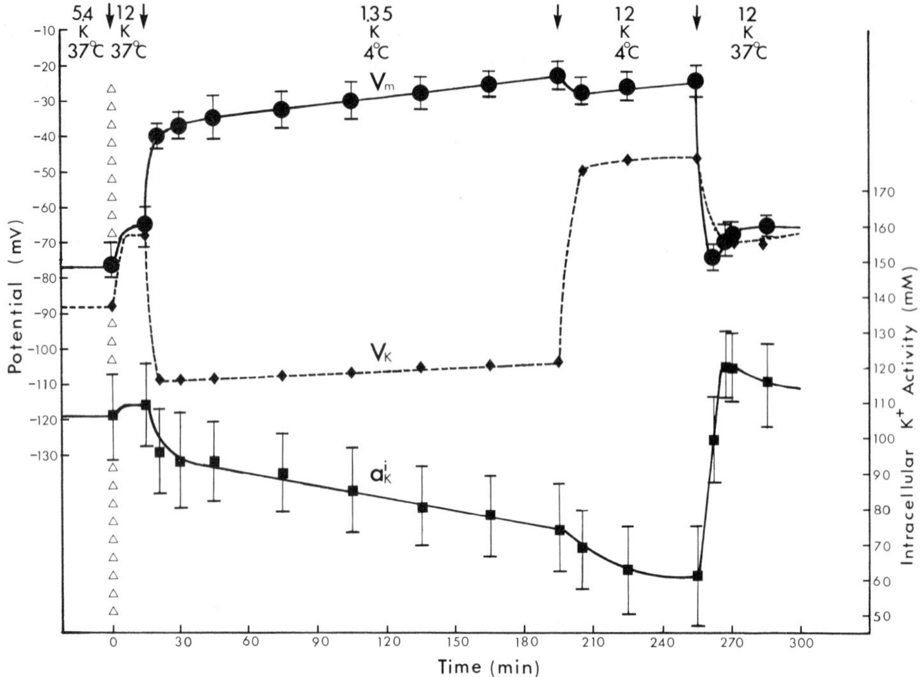

Fig. 3. The values of $V_m$, $V_K$ and $a_K^i$ during temperature and solution changes

## Discussion

It has been found that under physiological conditions the average value of $a_K^i$ in sheep cardiac Purkinje fibres is 110 mM (13). In this study, $a_K^i$ decreased to only 40% of its control value after 24 hours of cooling, indicating that either 4°C was insufficient to shut off the Na-K pump completely or that there were some other mechanisms available to maintain the transmembrane gradient of $a_K^i$ after pump inhibition. This phenomenon of changes in $a_K^i$ reaching a plateau level is clearly shown in Fig. 3, where $a_K^i$ decreased quickly during the first 10 - 15 minutes, but then slowly stabilized at a value much higher than that of the extracellular solution.

In analogous experiments Deitmer and Ellis (2) have shown, that pump inhibition by a high dose of cardioactive steroid produced an initially rapid increase of $a_{Na}^i$. After 15 - 30 minutes this $a_{Na}^i$ increase slowed down considerably and reached a plateau within 2 hours at a level much below the extracellular sodium activity. They proposed that the $Ca^{2+}$ gradient through the $Na^+-Ca^{2+}$ exchange mechanism was responsible for this constancy of $a_{Na}^i$; that is, intracellular $Na^+$ was exchanged for extracellular $Ca^{2+}$. Since $a_{Na}^i$ reached a stable value, $a_K^i$ might be expected also to reach a stable value. It is difficult to accept the $Ca^{2+}$ gradient as the basis for the maintenance of the other ionic gradients, since this would require a temperature-independent Ca pump. Further experiments will be needed in order to settle this question of $a_{Na}^i$ control in the cold.

The slight increase in $a_K^i$ in 12 mM $[K]_o$ at $37^oC$, confirms previous results (13). $V_m$ was quite close to $V_K$ in this solution at $37^oC$; however, this relation no longer held at $4^oC$. The membrane did not behave like a $K^+$-electrode at $4^oC$, even with a $[K]_o$ higher than normal. The augmentation of the rate of $a_K^i$ loss in 12 mM $[K]_o$ at $4^oC$ implies that after several hours of cooling the membrane permeability to $K^+$($P_K$) rather than the electrochemical gradient dominated the $a_K^i$ loss. In 1.35 mM $[K]_o$ at $4^oC$, $V_m-V_K$ was 60 mV larger than that in 12 mM $[K]_o$ at $4^oC$. The apparently small $P_K$ in the cooled fibres can explain the slow decrease in $a_K^i$. In addition, that the rate of $V_m$ depolarization was twice as large as that of $V_K$ indicated that both, the decrease of $a_K^i$ and the change in $P_K$ contributed to the depolarization.

The steady state pumping current in cardiac muscles under physiological conditions is about 3 $pmol/cm^2 \cdot s$ (5, 8). When the pump was stimulated by intracellular $Na^+$ loading, sodium efflux increased to 30 $pmol/cm^2 \cdot s$ (6). If the exchange ratio is 3 $Na^+$ to 2 $K^+$ (12), then the $K^+$ influx should increase to 20 $pmol/cm^2 \cdot s$. Our estimate of 25 $pmol/cm^2 \cdot s$ is quite close to this number. In addition, to the extent that there was a brief period with $V_m$ more negative than $V_K$, there should have been an inward $K^+$ leakage because of the reversal of the electrochemical gradient.

Finally, in this study, we observed a more rapid hyperpolarization transient during rewarming than that reported by Hiraoka and Hecht (7). The time course of changes in our experiments are closer to those observed by Gadsby and Cranefield (4).

References

1. Carmaliet E (1978) Cardiac transmembrane potentials and metabolism. Circ Res 42(5): 577
2. Deitmer JW, Ellis D (11978) The intracellular sodium activity of cardiac Purkinje fibres during inhibition and re-activation of the Na-K pump. J Physiol 284: 241
3. Fozzard HA, Lee CO (1976) Influence of changes in external potassium and chloride ions on membrane potential and intracellular potassium ion activity in rabbit ventricular muscle. J Physiol 256: 663
4. Gadsby DC, Cranefield PF (1979) Electrogenic sodium extrusion in cardiac Purkinje fibres. J Gen Physiol 73: 819
5. Glitsch HG (1979) Characteristics of active Na transport in intact cardiac cells. Am J Physiol 5(2): H189
6. Glitsch HG, Pusch H, Venetz K (1976) Effects of Na and K ions on the active Na transport in guinea-pig auricles. Pflügers Arch 365: 29
7. Hiraoka M, Hecht HH (1973) Recovery from hypothermia in cardiac Purkinje fibres: considerations for an electrogenic mechanism. Pfluegers Arch 339: 25

8. Isenberg G, Trautwein W (1974) The effect of dihydro-ouabain and lithium ions on the outward current in cardiac Purkinje fibres: evidence for electrogenicity of active transport. Pfluegers Arch 350: 41

9. Lee CO, Fozzard HA (1975) Activities of potassium and sodium ions in rabbit heart muscle. J Gen Physiol 65: 695

10. Mobley BA, Page E (1972) The surface area of sheep Purkinje fibres. J Physiol 220: 547

11. Page E, Storm SR (1965) Cat heart muscle in vitro VIII. Active transport of sodium in papillary muscles. J Gen Physiol 48: 957

12. Schwartz A, Lindenmayer GE, Allen JC (1975) The sodium-potassium adenosine triphosphatase: pharmacological, physiological and biochemical aspects. Pharmacol Rev 27: 3

13. Sheu SS, Korth M, Lathrop DA, Fozzard HA (1980) Intra- and extracellular K and Na activities and resting membrane potential in sheep cardiac Purkinje strands. Circ Res (in press)

14. Walker JL (1971) Ion specific-ion exchanger microelectrodes. Anal Chem 43: 89

# Intracellular pH in Purkinje Fibers. Effect of Extracellular Acidosis in a $CO_2/HCO_3$- and HEPES Containing Medium


A. de HEMPTINNE, R. MARRANNES


It is well documented that the mechanical performance of heart cells is very much influenced by the pH of the extracellular medium and is markedly depressed by acidosis. Experimentally it has been shown that the depression of the contractility of the cat papillary muscle is much more rapid when the induced acidosis is of the respiratory type (caused by an increase in the partial pressure of $CO_2$ at constant $HCO_3^-$ concentration) than when the acidosis is of the metabolic type (caused by a decrease in the $HCO_3^-$ concentration at constant partial pressure of $CO_2$ (5)).

Using pH selective microelectrodes in sheep cardiac Purkinje fibers, Ellis and Thomas (3) have shown that a respiratory type of extracellular acidosis causes a much more rapid intracellular acidification than the metabolic type. This difference is explained by the rapid penetration of $CO_2$ across the cell membrane causing after hydration and dissociation an intracellular acid load.

In metabolic acidosis, a relatively slow intracellular acidification was observed which was preceded by a transient change in the alkaline direction which the authors described as a regular finding of unknown origin.

The search for an explanation of the apparently unexpected transient change in intracellular pH under conditions of metabolic acidosis was the aim of this work.

## Methods

Sheep hearts were obtained from the local slaughterhouse. The heart was removed and immersed in a cold modified Tyrode solution which had been equilibrated with a gas mixture of 95% $O_2$, 5% $CO_2$. Purkinje fibers were isolated from the left ventricle and placed horizontally in an experimental chamber where they were continually superfused.

## Microelectrodes

In order to measure the intracellular pH, double barrelled pH sensitive microelectrodes were prepared as described previously (2). The reference barrel was filled with 3 M KCl. Some experiments were performed with single barrelled pH sensitive microelectrodes (6) using a separate KCl filled microelectrode to measure the membrane potential.

The surface pH was measured using a "recessed" type pH microelectrode whose tip was intentionally broken and fire polished to prevent cell damage when pressed onto the surface of a fiber.

In some experiments, a small bulb type electrode with a rounded tip of pH glass was used.

190

## Solutions

The normal superfusion solution equilibrated with nominally 95% $O_2$, 5% $CO_2$ contained (mmol/l) : NaCl 120.5; KCl 4; $CaCl_2$ 2.5; $MgCl_2$ 1.2; $NaHCO_3$ 25; $NaHPO_4$ 1.2; glucose 5.

Metabolic or chloride acidosis ($Cl^-$ ac) was obtained on replacing 20 mM $HCO_3^-$ with the same amount of $CL^-$. Organic acid acidosis was obtained on replacing 20 mM $CL^-$ from the $Cl^-$ ac solution with 20 mM organic anions. Respiratory acidosis was realized on equilibrating the normal perfusate with 75% $O_2$, 25% $CO_2$.

The $CO_2$-free solutions were saturated with 100% $O_2$, lacked $HCO_3^-$ but contained 10 mM of the Na salt of HEPES (N-2-hydroxyethyl piperazine-N'-2 ethanesulfonic acid (pKa 7.55)) and 10 mM Na salt of PIPES (Piperazine-N, N'-bis ethanesulfonic acid (pKa 6.8)).


## Results

a) Effect of extracellular acidosis on $pH_i$ in a $CO_2/HCO_3$ containing medium

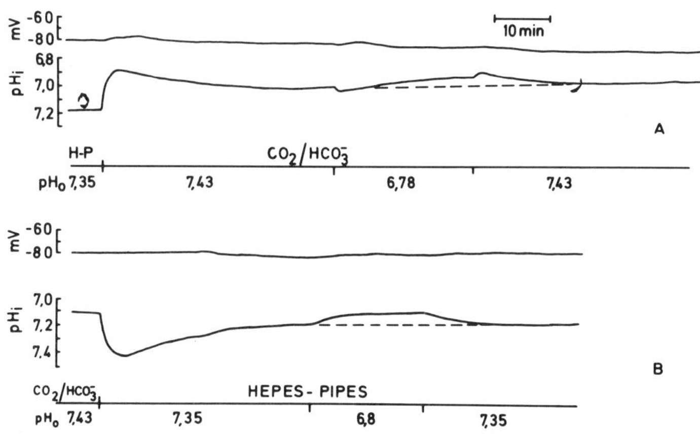


Fig. 1. Time course of change of membrane potential and intracellular pH. A. The superfusion solution, buffered initially with HEPES and PIPES (10 mmol/l each), was replaced by a $CO_2/HCO_3$ containing solution. The effect of extracellular acidosis ($pH_o$ = 6.78) is shown. B. The superfusion solution, initially buffered with $CO_2/HCO_3$, was replaced by a solution containing HEPES and PIPES. The effect of extracellular acidosis ($pH_o$ = 6.8) is shown

Figure 1 A illustrates the time course of change of $pH_i$ seen on switching first from a $CO_2$ free medium (buffered with HEPES and PIPES) to a $CO_2/HCO_3^-$ containing medium. The rapid intracellular acidification seen is classically explained by diffusion of $CO_2$ across the cell membrane causing, after hydration and subsequent dissociation, an intracellular acid load. An active pH regulatory system (possibly a proton pump) must be postulated to explain the slow shift of $pH_i$ to a more alkaline value after the initial acid change. When, at constant $pCO_2$, the superfusion solution is changed from pH 7.43 to 6.78, $pH_i$ shows a relatively slow acid shift preceded by a small transient change in the alkaline direction. The reverse can be seen on returning to the control solution. This confirms the observations made by Ellis and Thomas (3).

b) Effect of extracellular acidosis on $pH_i$ in a $CO_2$-free medium

Figure 1 B illustrates the time course of change of $pH_i$ seen on switching from a $CO_2/HCO_3^-$ containing medium to a $CO_2$-free medium (buffered with HEPES and PIPES). The rapid intracellular alkalinization seen is again classically explained by the rapid efflux of $CO_2$. When, thereafter, the superfusion solution is changed from pH 7.35 to 6.8, $pH_i$ shows an acid shift with no initial transients. On returning to the control solution (pH 7.35) $pH_i$ shifts to a more alkaline value following a single exponential time course.


Discussion

Previous work concerning the effect of organic acids on the intracellular and surface pH has led us to develop a three compartment model (the bulk solution, the interstitial or surface compartment and the intracellular compartment) in order to explain the observed pH changes (4). This model can be described mathematically when one assumes that the diffusion of both the dissociated and undissociated form of an organic acid between the bulk and the interstitial compartment can be described by the Fick equation with an equal permeability constant. The diffusion of the undissociated form of the acid across the cell membrane can also be assumed to obey Fick's law, while the diffusion of the dissociated form can be described by the constant field equation as suggested by Boron and Deweer (1). An active proton extrusion mechanism has to be incorporated into the model in order to account for the steady state value of $pH_i$ which is much more alkaline than the theoretical electrochemical equilibrium value for the protons. In Boron and Deweer's model, a mathematical expression is used which accounts for an additional pumping rate in response to an intracellular acid load.

However, in order to explain the intracellular acid shift in the steady state condition which is seen when the extracellular medium is made acid in a $CO_2$-free medium (cf. Fig. 1), we assume that the pumping rate of protons is negatively influenced by an extracellular acid load. In the formulation that we have chosen, the pumping rate is stimulated by intracellular acidosis and depressed or even transiently reversed by extracellular acidosis. Reverse pumping is assumed to occur when a large inwardly directed gradient exists between the extracellular and intracellular proton concentration. The details of the mathematical formulation are to be published elsewhere. Fig. 2 gives an illustration of the calculation of $pH_i$ from the model on simulating the experimental conditions given in Fig. 1. The calculated curve of $pH_i$ reproduces the transient alkaline change when the bulk solution buffered with

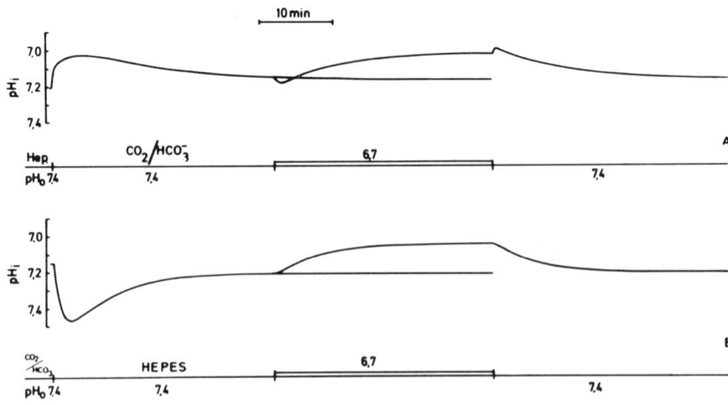

Fig. 2.  Time course of change of intracellular $pH_i$ as predicted by a mathematical model which simulates the experimental conditions given in Fig. 1

$CO_2/HCO_3^-$ is made more acid by reducing the $HCO_3^-$ concentration at constant $CO_2$ pressure. The transient acid change of $pH_i$ is also reproduced on returning to the control solution. The transient changes of $pH_i$ are absent when the same extracellular pH changes are tested in a simulated $CO_2$ free condition, the bulk solution being then buffered with HEPES and PIPES which do not penetrate into the cells. From the three compartment model schematically reproduced in Fig. 3, the transient intracellular shift in the alkaline direction seen when the $HCO_3^-$ concentration of the bulk solution is suddenly reduced to 5 mM at constant partial pressure of $CO_2$ can be explained as follows. As a result of the concentration gradient of ions between the bulk solution and the interstitial compartment, $HCO_3^-$ will diffuse from the interstitial space to the bulk solution and protons will diffuse in the reverse direction. In the interstitial space the overall direction of the reaction "$CO_2 + H_2O \rightleftharpoons H_2CO_3 \rightleftharpoons H^+ + HCO_3^-$" depends on variations in the product $(a_{H^+} \cdot a_{HCO_3^-})$. This product, calculated for the interstitial space, does not undergo large changes when the bulk solution is made more acid by decreasing $HCO_3^-$ because the percent decrease in $HCO_3^-$ in the interstitial space should be compensated by a similar percent increase in $H^+$ concentration as a result of its diffusion. This is so provided  no extra buffer capacity (besides the diffusible buffers) is attributed to that space. When however, the interstitial space is assumed to possess some extra buffer capacity ( $\delta_s$, due to the presence of protonatable, fixed negative charges located at the surface of the cells), the given product will tend to decrease transiently in metabolic acidosis. This can be expected because the drop in $HCO_3^-$ concentration is less compensated for by a simultaneous percent increase in $H^+$ concentration as the protons which penetrate into the space can be buffered by the

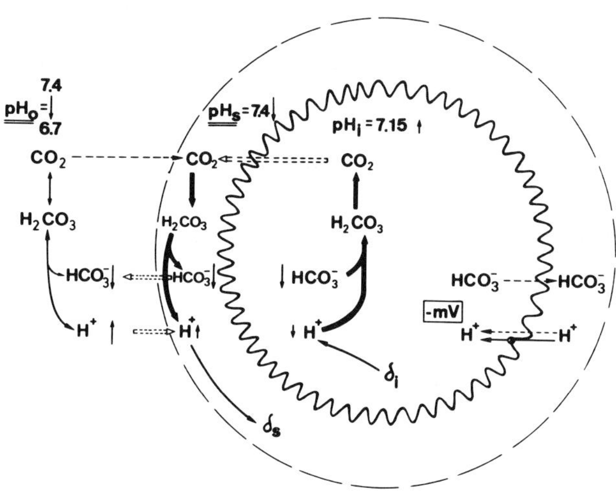

Fig. 3. Schematic representation of a three compartment model. The cell
membrane is represented by the inner circle with many infoldings. The interstitial space is surrounded by the outer circle which acts as a diffusion
limiting barrier. The horizontal broken arrows show the initial direction
of passive diffusion of $CO_2$, $HCO_3^-$ and $H^+$ when the pH of the bulk solution
($pH_o$) is changed from 7.4 to 6.7 at constant $CO_2$ pressure. The uninterrupted
horizontal arrow gives the direction of transport by the proton pump. Reverse pumping is assumed to take place. The intracellular and interstitial,
non $CO_2/HCO_3$ dependent, buffer capacities are represented by $\delta_i$ and $\delta_s$. More
explanation is to be found in the text

local non diffusible buffers. When the value of the product ($a_{H^+} \cdot a_{HCO_3^-}$)
decreases, the reaction given above will proceed to the right and cause a
transient decrease of $CO_2$ pressure in the interstitial space. The local $CO_2$
sink thus obtained attracts $CO_2$ from the two neighbouring compartments :
the bulk solution and the cellular compartment. When $CO_2$ leaves the cells,
the above mentioned reaction will proceed to the left within the cells producing a transient intracellular alkalosis.

References


1. Boron WF, De Weer P (1976) Intracellular pH transients in squid giant
   axons caused by $CO_2$, $NH_3$ and metabolic inhibitors. J Gen Physiol 67:
   91–112
2. de Hemptinne A (1979) A double-barrel pH microelectrode for intracellular
   use. J Physiol (Lond) 295:5–6P

194

3. Ellis D, Thomas RC (1976) Direct measurement of the intracellular pH of
      mammalian cardiac muscle. J Physiol (Lond) 262:755-771
4. Marrannes R, de Hemptinne A, Leusen I (1979) Correlation between con-
      duction velocity transients in isolated heart fibers and pH changes
      (interstitial and intracellular). Arch Int Physiol Biochim 87:770-772
5. Pannier JL, Leusen I (1968) Contraction characteristics of papillary
      muscle during changes in acid-base composition of the bathing fluid.
      Arch Int Physiol Biochim 76:624-634
6. Thomas RC (1974) Intracellular pH of snail neurones measured with a new
      pH-sensitive glass microelectrode. J Physiol (Lond) 238:159-180


Laboratory of Normal and Pathological Physiology, University of Gent,
9000 Gent, Belgium


## Discussion

Ammann:  The speed of response of recessed tip glass pH-microelectrodes
seems to be inadequate for certain studies of pH changes. Liquid membrane
microelectrodes generally show much shorter response times than the recessed tip microelectrodes. The figure demonstrates the state of the art in a
liquid membrane electrode based on a lipophilic, synthetic $H^+$-carrier (membrane composition see figure; Simon et al. Proc. Analytiktreffen (1979),
Chemische Ges. der DDR (Ed.) (1980), in press). The corresponding PVC-membrane electrodes have been described before (Erne et al. Chimia (Switzerland) 33:88 (1979)). The figure shows that in TRIS buffered solutions containing a representative extracellular ion-background a useful response to
$H^+$-activity changes, in the physiological pH range, is obtained. The same results were observed for a typical intracellular ion-background ($Na^+$: 10 mM;
$K^+$: 200 mM; $Ca^{++}$: 0.01 MM). So far, the limiting factor in view of physiological applications of this liquid membrane electrode is the rather high
membrane resistance (tip diameter 2.5 um: $10^{12}\Omega$). Work is in progress to improve the microelectrode, especially with respect to membrane resistance.

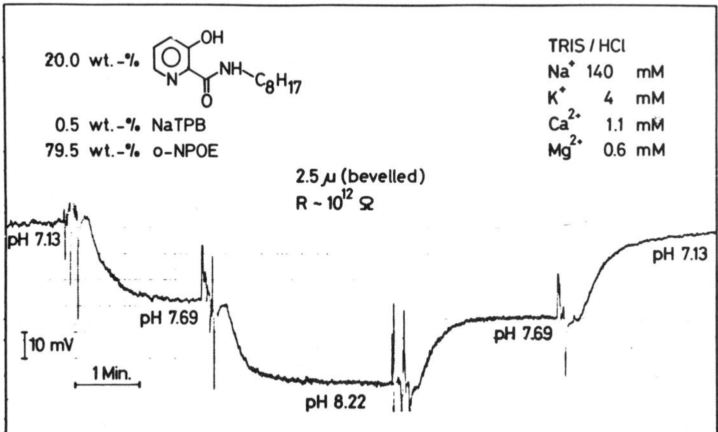


State of the art in a liquid membrane microelectrode based on a lipophilic,
synthetic $H^+$-carrier

# Intracellular Electro-Chemical Studies of Single Renal Tubule Cells and Muscle Fibers

R.N. KHURI, S.K. AGULIAN

## I. Intracellular Electrochemistry

The true internal environment is the cytoplasmic aqueous solution each cell
contains within its membrane. The cytoplasmic aqueous solution is known as
the cytosol. Knowledge of intracellular electrolyte composition is essential
for the understanding of cell function. Certainly the more important changes
occur inside cells.

In 1949 Ling and Gerard (29) reported the first direct intracellular electrical measurement. In 1954 Caldwell (2) reported the first direct intracellular electrometric determination of a cytoplasmic ionic constituent,
the intracellular pH of giant crustacean muscle fibers. In 1959 Hinke (7)
determined the intracellular $K^+$ and $Na^+$ of squid giant axon using microelectrodes made of cation-selective glasses. In 1964 Lev (28) and in 1969
Kostyuk et al. (27) reported the results of determinations of intracellular
pH, $Na^+$ and $K^+$ in single muscle fibers of the frog sartorius by means of
glass microelectrodes. In all these studies, the electrical and chemical
(ionic) determinations were made by means of separate probes.

The combination of the electrical and the chemical microprobes in a double-
barrelled configuration for the simultaneous monitoring of the intracellular
electrical and chemical ionic potentials was first achieved by Khuri and co-
workers (11, 12, 23, 24). Liquid ion-exchangers were employed as the chemical ionic sensors. The advantages of liquid ion-exchangers as the sensing
elements of ion-selective microelectrodes are: 1) the ease with which
double-barrelled micropipettes can be fabricated, 2) the degree of minia-
turization that can be achieved, and 3) the fast response time.

The improved method of construction of double-barrelled microelectrodes
over that reported by Khuri et al. (23, 20) is described below. This improved method is characterized by ease of manufacture and high yield of
functional microelectrodes.

Pyrex capillary tubing (Corning Code 7740) with an o.d. 1.2 mm and i.d. 0.6
mm is cleaned with dichromic acid, distilled water, and acetone. The capillaries are dried in the oven. Then a segment of pyrex tubing 10 cm in length
is attached to a borosilicate glass capillary with inner filament (WP-Instruments, Cat. No. IB120F), o.d. 1.2 mm and i.d. 0.68 mm, by means of shrinkable tubing (Alpha Wire Corporation, fit 105 3/32"), heated on a microflame.
The ends of the glass capillaries are fire polished. The positions of the
shrinkable tubings are shown in Fig. 1 (upper) as x, y & z.

The pair of tubings are held over a microflame between y & z and when sof-
tened rotated 360° and pulled slightly. Then it is mounted on a vertical pipette puller (David Kopf Instrument) and pulled into double-barrelled micropipettes with relatively long shanks and tip diameter $<1\,\mu$. The lower micropipette is used only (Fig. 1 lower).

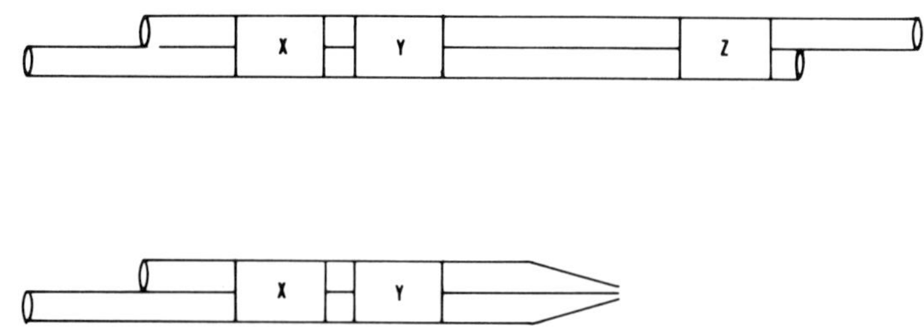


Fig. 1. Stages in fabricating electrodes. Upper: Position of shrinkable
tubing. Lower: Simultaneously pulled micropipettes

To ensure best electrode performance and maximum yield the sequence in
electrode filling is strictly followed. First, the indicator barrel is
filled by introducing a column of liquid ion exchanger into its shank by
means of a filler capillary. Pressure is applied to the column until it is
pushed as far as to the tip of the micropipette. The rest of the shank and
the stem of the indicator barrel are then filled with 3 M NaCl as an internal reference solution.

Second, the reference barrel is filled with an appropriate salt bridge electrolyte (3 M NaCl, KCl and Na formate are used for $K^+$, $Na^+$ and $Cl^-$ measurements respectively), by wetting the inner filament of the borosilicate glass
capillary. Filling is performed under microscopic inspection to ensure that
there is no air bubble trapped in the tip of the reference barrel.However,
if an air bubble is trapped in the shank it is removed by applying and releasing pressure by means of a syringe. The rest of the shank and the stem
are then filled with a filler capillary. A piece of PE tubing is fitted to
the stems of both reference and indicator barrels and are filled with appropriate internal reference solutions, as are mentioned above. Fig. 2 is
a diagrammatic representation of the electrode.

The combination electrode was mounted on an electrode carrier connected to
a hydraulic micro-drive (David Kopf, model 1207S). The electrode carrier
itself is mounted on a Leitz micro-manipulator. An Ag-AgCl wire is inserted into the stem of each barrel as an internal reference element. The
indicator barrel is tightly fitted to a lucite chamber, as shown in Fig. 2,
filled with internal reference solution. Pressure is applied to the indicator barrel in order to retain the organic ion-exchanger within the con-
fines of the tip. In this manner the aqueous phase is prevented from enter-
ing the terminal part of the micropipette shank through the single porous
channel at the tip. Leakage of organic exchanger from the tip of the electrode or manifestation of electrical coupling between the two barrels is

indicative of broken electrode tips. The leads of the electrode and the external salt bridge reference are connected to the inputs of a differential/ dual high impedance electrometer (WP-Instruments, Inc. Model F223A). The readings of the potential measurements were displayed on a grass polygraph. Isothermal conditions were maintained at $25 \pm 1^{\circ}$C. All equipment was placed inside a radio frequency shielded room.

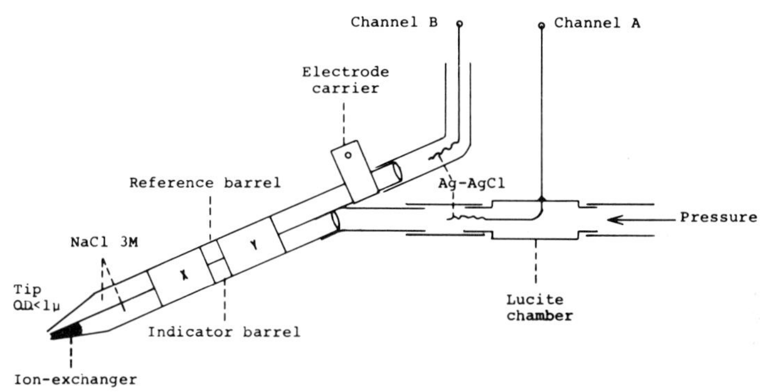


Fig. 2.   Improved double-barrelled liquid ion-exchange microelectrode

Freshly prepared electrodes are immediately tested for any electrical coupling between the two barrels in the region of the tip. This is accomplished by measuring the electrical potential between the reference of the combination electrode and an external reference electrode in solutions with varying ion concentrations. A constant negative potential of few millivolts is indicative of no interference. Simultaneously, the sensitivity of the electrode is determined by measuring the potential between the indicator barrel and the external reference electrode. Electrodes exhibiting a response of 55-60 mV/decade change of ion activity and a response time of less than one second are selected for kinetic studies.

Fig. 3 represents a simultaneous steady-state electrochemical recording of the intracellular electrical (upper tracing) and ionic potentials. The electrical component ($V_m$) of the lower tracing can be subtracted automatically by using the differential operator of a dual electrometer, thus leaving a pure ionic potential. As shown in the figure, the total response time (electrometric + mechanical + biological) is only 5 sec for the potential to attain 95% of the new steady-state value.

The intracellular electrochemical technique is particularly useful for monitoring and comparing the time course of the transient electrical and chemical potentials induced by rapid changes in extracellular perfusion fluids. Fig. 4 is an example of such a kinetic tracing. With a double-barrelled electrode continuously recording cell membrane electrical PD($V_m$) and cytosolic $Na^+$ activity, two perfusion solutions are rapidly switched over

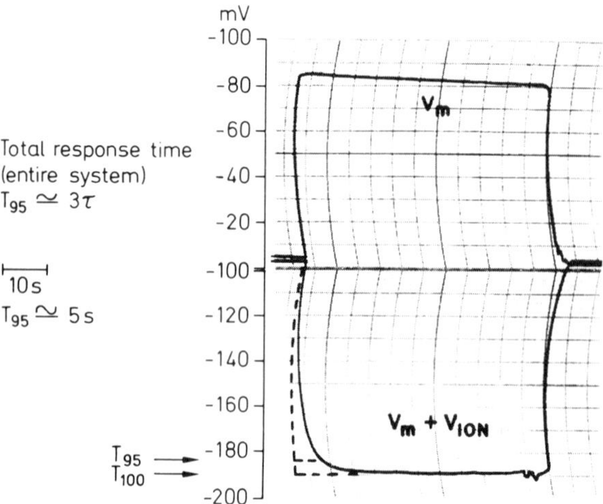



Fig. 3.  Simultaneous electrochemical recording of electrical and ionic potentials, as double-barrelled microelectrode impales cell membrane and achieves stable intracellular localization of its tip

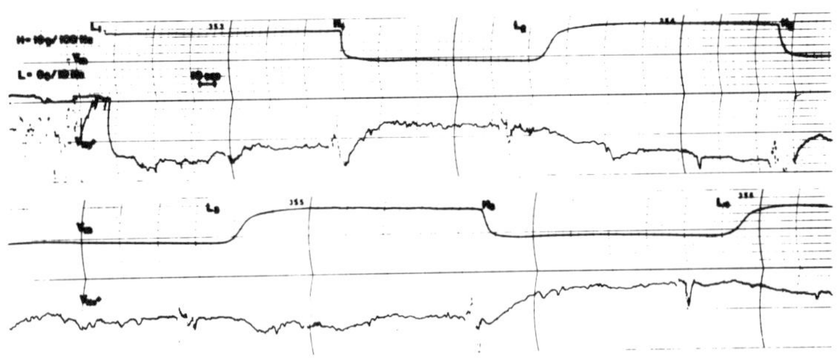

Fig. 4.  Continuous monitoring from within a single cell of reversible changes in intracellular electrical and sodium potentials induced by instantaneous changes in composition of two luminal perfusion fluids

several cycles. The induced changes are reversible such that each cell serves as its own control. The rate of change of the ionic potential is generally slower by an order of magnitude than the rate of change of the associated electrical potential. Several parameters that characterize the kinetics of the ionic permeation of a given cellular membrane may be obtained from tracings as in this figure. These include ionic transference numbers, ionic fluxes and ionic permeabilities. The latter two parameters are calculated from the initial slope of the change in cytosolic ionic activity.

<u>II. Potassium</u>

The activity of potassium in living protoplasm plays a major role in the electrochemistry of cells in general and excitable cells in particular. Studies in skeletal muscle, employing different methods, have generally concluded that all the intracellular $K^+$ is free and evenly distributed within the entire myoplasm of the muscle fiber. Potassium is the major intracellular cation.

Skeletal muscle constitutes the major fraction of the total body mass. Therefore, muscle tissue is the major determinant of intracellular fluid ionic composition. Muscle can serve as a general biological reference for making comparisons of intracellular ionic composition.

The $K^+$-selective liquid ion-exchanger, a 1% substituted tetraphenylboriole salt in 3-nitro-O-xylene (Corning Catalog No. 476132) is an organic electrolyte dissolved in an organic solvent. A column of the potassium exchanger is injected into the tip of the indicator barrel.

Khuri et al. (25) found an electrometric intracellular $[K^+]$ of 163.2 $\pm$ 3.7 mM in single fibers of the gastrocnemius muscle of the rat under normal control conditions. This value is in close agreement with the total apparent $[K^+]$ as derived from chemical analysis (9, 33), indicating that all the intracellular $K^+$ is free and evenly distributed within the entire myoplasm. Intracellular $[K^+]$ drops to 123.9 $\pm$ 2.3 mM on a chronic low potassium diet and to 132.7 $\pm$ 3.8 mM with acute acidosis. The calculated $K^+$ equilibrium potential ($E_K$) was under all studied conditions in excess of 10 mV more electronegative than the simultaneously measured membrane resting potential ($E_m$). This positive electrochemical disequilibrium of the potassium ion can be taken as evidence for active $K^+$ influx by a coupled Na-K exchange pump.

The <u>Necturus</u> kidney proximal tubule offers a combination of large cells and absence of pulsations, features that render electrometric analysis relatively simple. It is particularly well suited to kinetic analysis where continuous intracellular monitoring is required. In 1972 Khuri et al. (24) reported a mean intracellular $K^+$ activity of 58.7 $\pm$ 2.3 mM, a value which was confirmed in 1977 by Fujimoto et al. (5) who obtained a $K^+$ activity of 61.1 $\pm$ 1.8 mM in the bullfrog proximal tubule. From this and the apparent total chemical $K^+$ concentration of 103 mM, an apparent cytosolic $K^+$ activity coefficient of 0.57 was obtained. In a subsequent series of experiments with <u>Necturi</u> (18) an even lower cytosolic $K^+$ activity coefficient of 0.46 was obtained. The low cytosolic $K^+$ activity and activity coefficient indicates that some 25 - 40 percent of the cell $K^+$ is either bound or sequestered in potassium-rich subcellular organelles. Had all the $K^+$ content of the <u>Necturus</u> proximal tubule cell been in free solution in its cytosol it would have a $K^+$ equilibrium potential ($E_K$) of about 95 mV, a value which is comparable to muscle. However, the reduction of intracellular $K^+$ activity lowers $E_K$ by 20 mV to a value of 75 mV. Although close to $E_m$, this value of $E_K$ remains some 5 - 7 mV more electronegative than $E_m$. Thus $K^+$ influx across the peritubular membrane must be uphill, but this active transport process is overcoming a rather small electrochemical gradient.

The distal tubule is the major determinant of $K^+$ secretion and, therefore, excretion in the final urine. Besides sharing with other amphibian kidneys the advantage of not being pulsatile, the <u>Amphiuma</u> kidney (30) has a ventral surface composed almost exclusively of distal tubules. Khuri et al. (19) found a mean value of $K^+$ activity of 47.0 $\pm$ 2.0 mM which, when taken in conjunction with the chemical $K^+$ concentration, yields an apparent activity coefficient for cell $K^+$ of 0.46. This suggests that some 40% of cell $K^+$ is sequestered and/or bound. Kimura et al. (26), using single-barrel $K^+$ microelectrodes found an intracellular $K^+$ activity of 41.2 $\pm$ 0.5 mM in the toad urinary bladder. This value corresponds to an apparent cell $K^+$ activity coefficient of 0.32. White (34) obtained an intracellular $K^+$ activity of 41.6 $\pm$ 1.5 mM in the absorptive cells lining the small intestine of Amphiuma. This corresponds to an apparent cell $K^+$ activity coefficient of 0.28.


## III. Sodium

It is generally accepted that the major portion of cytoplasmic sodium is not free. Since active $Na^+$ transport is a universal function of virtually all living cells, the low intracellular $Na^+$ maintains a favorable electrochemical gradient for passive $Na^+$ influx. This steep gradient is particularly useful as a driving force for the action potential of excitable tissues.

Several studies of intracellular $Na^+$ of muscle have been carried out with the aid of $Na^+$-sensitive glass microelectrodes in large muscle fibers. Hinke (7) obtained an activity coefficient of about 1/3 of that in extracellular fluid in giant crab muscle fibers. Lev (28) and Kostyuk et al. (27) obtained similar values in frog sartorius fibers.

Khuri in the first reported use of a $Na^+$-selective liquid ion-exchange microelectrodes found a mean myoplasmic $Na^+$ activity of 4.4 $\pm$ 0.3 mM in the gastrocnemius muscle of the rat in vivo. The $Na^+$ liquid ion-exchanger is a neutral $Na^+$ ligand developed by W. Simon and reported by Steiner et al. (33). The above $Na^+$ activity would yield an apparent activity coefficient of 0.27, indicating that some 64 percent of myoplasmic $Na^+$ is bound and/or compartmentalized. The $Na^+$ may be bound to the muscle protein myosin. The intracellular organelles capable of $Na^+$ sequestration include the nucleus, the mitochondria, and the sarcoplasmic reticulum.

In rat muscle it is apparent that the two major intracellular monovalent cations, $K^+$ and $Na^+$, differ markedly in their intracellular physical states. While virtually all the cell $K^+$ is in free solution, most of the cell $Na^+$ is not, suggesting that the intracellular processes of sequestration and/or binding have a preferential affinity for the $Na^+$ over the $K^+$ ion.

In the first direct electrometric determination of intracellular $Na^+$ activity of renal tubular cells, Khuri et al. (18) obtained a mean value of 20.0 $\pm$ 1.3 mM in the proximal tubule of Necturus. This means that some 60 percent of renal cell $Na^+$ is free and 40 percent bound and/or sequestered. The apparent activity coefficient is 0.48. Thus the intracellular $Na^+$ and $K^+$ activity coefficients in cells of the proximal tubule of Necturus are quite similar, indicating no preferential binding and/or sequestration of the sodium ion as in muscle.

In order to obtain direct evidence for our understanding of the luminal $Na^+$ entry, step experiments with different luminal perfusions were carried out. Increasing luminal perfusion fluid Na from 10 to 100 mM resulted in an increase in cell ($Na^+$) of 13.3 $\pm$ 2.5 mM and an increase in cell ($K^+$) of 12.5 $\pm$ 1.1 mM. This parallel increase in $Na^+$ and $K^+$ is due to the coupled

$Na^+$-$K^+$ exchange pump in the peritubular membrane. Increasing luminal perfusion fluid from zero (10 mM sucrose) to 10 mM glucose resulted in an increase in cell ($Na^+$) of 11.6 $\pm$ 1.6 mM. This lends evidence to the presence of luminal membrane co-transport of $Na^+$ and glucose and suggests a glucose/$Na^+$ transport stoichiometric ratio of 1:1. To study the dependence of $Na^+$ transport on the lumen/cell $Cl^-$ gradient the $Cl^-$ of the luminal perfusion fluid was dropped from 98.1 to 5.6 mM. As a result cell ($Na^+$) fell by 10.6 $\pm$ 1.2 mM. This observation lends evidence for the mechanism of co-transport of $Na^+$ and $Cl^-$ as NaCl across the luminal membrane. Lowering the luminal perfusion fluid pH from 7.5 to 6.0 resulted in a fall in cell ($Na^+$) activity of 8.0 $\pm$ 1.2 mM. This fall in cell $Na^+$ with luminal acidification can be explained by the inhibition of luminal Na-H exchange. The co-transport of $Na^+$ and $Cl^-$ and of $Na^+$ and glucose and the countertransport of $Na^+$ and $H^+$ in the luminal membrane are depicted in the proximal tubule cell scheme of Fig. 5.

In an effort to define the properties of the peritubular membrane, the $[K^+]$ of the peritubular perfusion fluid was raised by the substitution of KCl for NaCl. As the peritubular perfusate $[K^+]$ is increased from the control value of 2.5 mM to 103 mM the cytosolic ($K^+$) increased while the cytosolic ($Na^+$) decreased by almost equivalent amounts. These equivalent but opposing changes in the cell ($K^+$) and ($Na^+$) constitute evidence for a coupled $Na^+$-$K^+$ exchange pump in the peritubular membrane (as represented in Fig. 5).

The distal tubular $Na^+$ reabsorptive system is responsible for the fine adjustment of renal $Na^+$ regulation. Khuri et al. (19) obtained a mean cytosolic ($Na^+$) activity of 16.3 mM in an electrometric study of _Amphiuma_ distal tubules. This would yield an apparent activity coefficient of 0.45 and suggests that 60% of cell $Na^+$ is in free solution in the distal tubule of _Amphiuma_ as in the proximal tubule of _Necturus_. Thus $Na^+$ and $K^+$ in the distal tubule of _Amphiuma_ have the same apparent activity coefficients and are both free to the same extent, i.e. 60%. An electrometric study of $Na^+$ in the distal tubule of the rat by Khuri et al. (14) gave a mean cytosolic $Na^+$ activity of 16.4 $\pm$ 1.9 mM, a value which is in close agreement with that in _Amphiuma_.


## IV. Chloride


In 1941 Boyle and Conway (1) reported that the chloride distribution across the sarcolemma of skeletal muscle fibers is passive and reflects a Donnan-type distribution. In 1974 a student of Conway, Kernan (10) challenged the passive distribution theory of $Cl^-$ in muscle with electrometric results obtained with chlorided silver wire microelectrodes. However, using $Cl^-$-selective liquid ion-exchanger microelectrodes, Khuri et al. (21) found that in the gastrocnemius muscle of the rat in vivo the measured electrical PD ($E_m$) and the transcellular $Cl^-$ equilibrium potential ($E_{Cl}$) are virtually identical, suggesting that in skeletal muscle $Cl^-$ is in a state of passive electrochemical equilibrium distribution.

It was generally assumed that $Cl^-$ transport in both the proximal tubule (3, 6) and the distal tubule (4, 8) is passive. This conclusion was based on studies involving the transepithelial distribution of $Cl^-$ and other methods. However, intracellular $Cl^-$ electrochemical analysis in the proximal tubule of _Necturus_ and the distal tubule of the rat by Khuri et al. in 1974 and 1975 (15, 16) yield a mean cell $Cl^-$ of 18.7 $\pm$ 1.3 mM in the former and

42.3 $\pm$ 3.1 mM in the latter. These relatively high cytosolic $Cl^-$ values place intracellular $Cl^-$ at a higher electrochemical potential than either luminal or peritubular fluid. This will drive passive efflux of $Cl^-$ across both cell membranes. However, for tubular $Cl^-$ absorption to occur an active chloride mechanism must exist in the luminal membrane. This could be either a $Cl^-$ anionic pump or an electrically neutral NaCl cotransport mechanism.

Thus in both the proximal and distal tubules the $Cl^-$ reabsorptive process consists of two steps in series: primary ($Cl^-$ anionic pump) or secondary (cotransport of NaCl) active $Cl^-$ transport from lumen to cell, and passive diffusion of $Cl^-$ from cell to interstitium. In contrast, Na transport is passive across the apical membrane and active across the basal membrane.


## V. Bicarbonate

A bicarbonate-selective liquid ion-exchanger consisting of a 3:1:6 mixture of tri-n-octylpropylammonium chloride to octanol to tri-fluoracetyl-$\underline{p}$-butyl benzyne was first used as a sensor to determine cytosolic $[HCO_3^-]$ in muscle and proximal tubule.

Intracellular $[HCO_3^-]$ was determined in vivo in frog sartorius and rat gastrocnemius muscles (22). The mean values were 4.4 $\pm$ 0.3 mM in frog and 12.6 $\pm$ 0.6 mM in rat, both values are an order of magnitude greater than predicted for simple passive distribution. This suggests that either $HCO_3^-$ is actively transported into the muscle fiber or that $H^+$ is actively extruded out of the fiber. From these intracellular $[HCO_3^-]$ values one can calculate a myoplasmic pH of 7.00 for frog muscle and 7.14 for rat muscle, both being alkaline disequilibrium pHs. Obviously a high intracellular $[HCO_3^-]$ would make a large contribution to the buffering capacity of a muscle fiber.

As yet there has been no electrometric determinations of renal cell pH. This determination awaits the development of $H^+$-sensitive liquid ion-exchangers. In 1974 Khuri et al. (17) reported a mean value of cytosolic $[HCO_3^-]$ in single cells of the proximal tubule of <u>Necturus</u> of 11.1 $\pm$ 0.6 mM, a value which is well above 0.7 mM predicted for passive distribution. Similarly, a high intracellular $[HCO_3^-]$ of 25.5 $\pm$ 1.1 mM was obtained in the late proximal convoluted tubule of the rat kidney (13). Again, the high measured intracellular $[HCO_3^-]$ on the alkaline side of equilibrium, and the corresponding calculated intracellular pH of 7.43 $\pm$ 0.02, a value which is 1.3 pH units greater than the equilibrium value. In the studies where the $[Na^+]$ of luminal perfusion fluid was increased, while the perfusate bicarbonate concentration was kept constant, there was an accompanying increase of intracellular $[HCO_3^-]$ of 8 mM. This finding is evidence for a luminal $Na^+$-$H^+$ exchange mechanism. However, the high steady-state levels of intracellular $HCO_3^-$ and pH are consistent with the two hypotheses (32) advanced as the underlying mechanism of renal tubular acidification: $H^+$ secretion and $HCO_3^-$ absorption across the luminal membrane.


## VI. Cellular ionogram

Fig. 5 is a <u>Necturus</u> proximal tubule cell model. The bracketed [ ] values of the four major monovalent intracellular ions are concentrations. Conclusions may be based on a collective ionic framework.

Evaluation of cytosolic electroneutrality conditions reveals an apparent anion deficit of about 70 mEq $[(75K^+ + 25Na^+) - (19Cl^- + 11HCO_3^-) = 70]$. Since this anionic deficit is filled by polyvalent anions, the cytosolic solution is hyperionic, i.e., has a higher ionic strength than plasma. Hence, one should not assume an identity of ionic activity coefficients between extracellular and intracellular fluids. This is an additional advantage for the electrometric determination of ionic activities of intracellular fluid.

In Fig. 5 the proximal tubule cell schema is based on the total direct evidence obtained from the electrochemical studies. The primary active event is the operation of the peritubular active $Na^+ - K^+$ exchange pump. The $Na^+$ extrusion generates a favorable drive for passive luminal $Na^+$ influx. This transluminal $Na^+$ electrochemical gradient of 93 mV may provide a sufficient driving force to drive several ionic species uphill into the cell. These include co-transport of $Na^+$ with $Cl^-$ or even $HCO_3^-$. The latter is indistinguishable from the countertransport of $Na^+$ and $H^+$. They also include the co-transport of $Na^+$ with glucose.

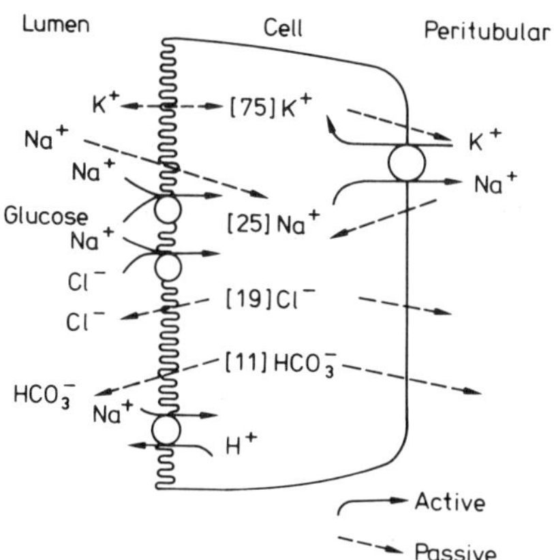


Fig. 5. Scheme of Necturus proximal tubule cell

By virtue of direct and accurate determinations of intracellular ionic activity, it was concluded that renal epithelium differs from skeletal muscle in at least three ways with regard to the chemical potential of its cytoplasmic monovalent ions. While intracellular $K^+$ is all free in muscle, some 40% of the renal cell $K^+$ is sequestered in subcellular organelles and not in free solution in the cytosol ($K^+$ activity coefficient of 0.5). While intracellular $Na^+$ is largely not free in muscle, at least 60% of renal cell $Na^+$ is in free solution in the cytosol ($Na^+$ activity coefficient of 0.5). While $Cl^-$ ion is passively distributed across the muscle fiber membrane, the high intracellular $Cl^-$ of renal cells requires the operation of some active mechanism (primary or secondary active transport).

In addition the study of the electrochemistry of the nephron has provided direct evidence in support of the 'double membrane' theory of epithelia and in support of the general membrane theory.

References


1. Boyle PJ, Conway EJ (1941) Potassium accumulation in muscle and associated changes. J Physiol 100:1
2. Caldwell PC (1954) An investigation of the intracellular pH of crab muscle fibers by means of microglass and micro-tungsten electrodes. J Physiol 126:169
3. Danielson BG, Persson E, Ulfendahl HR (1970a) Transmembrane transport of chloride and iodide in proximal rat tubules. Acta Physiol Scand 78:339
4. Danielson BG, Persson, E, Ulfendahl HR (1970b) The transport of halide ions across the membrane of distal rat tubules. Acta Physiol Scand 78:347
5. Fujimoto M, Kubota T, Kotera K (1977) Electrochemical profile of K and Cl ions across the proximal tubule of bullfrog kidneys. Contrib Nephrol 6:114-123
6. Giebisch G, Windhager EE (1964) Renal tubular transfer of sodium, chloride and potassium. Am J Med 36: 643-669
7. Hinke JAM (1959) Glass microelectrodes for measuring intracellular activities of sodium and potassium. Nature (London) 184:257
8. Kashgarian MH, Stöckle H, Gottschalk CW, Ullrich JJ (1963) Transtubular electrochemical potentials of sodium and chloride in proximal and distal renal tubule during antidiuresis and water diuresis (diabetes insipidus). Pfluegers Arch 277:89
9. Kendig JJ, Bunker JP (1970) Extracellular space, electrolyte distribution and resting potential in K depletion. Am J Physiol 218:1737
10. Kernan RP, MacDermott M, Wesphal W (1974) Measurement of chloride activity within frog sartorius muscle fibers by means of chloride-sensitive microelectrodes. J Physiol 241:60P
11. Khuri RN (1971) Intracellular potassium and the electrochemical properties of striated muscle fibers. Proc IUPS 9:301
12. Khuri RN (1972) Intracellular potassium in cells of the distal tubule. Yale J Biol Med 45:384
13. Khuri RN, Agulian SK (submitted) Intracellular bicarbonate in single cells of proximal tubule of rat. J Appl Physiol
14. Khuri RN, Agulian SK, AbdelNour S (submitted) Intracellular activity of sodium in the distal tubule of the rat. Pfluegers Arch
15. Khuri RN, Agulian SK, Bogharian K (1974) Electrochemical potentials of chloride in distal renal tubule of the rat. Am J Physiol 227:1352
16. Khuri RN, Agulian SK, Bogharian K, Aklanjian D (1975) Electrochemical potentials of chloride in proximal renal tubule of Necturus maculosus. Comp Biochem Physiol A 50:695
17. Khuri RN, Agulian SK, Bogharian K, Nassar R, Wise W (1974) Intracellular bicarbonate in single cells of Necturus kidney proximal tubule. Pfluegers Arch 349:295-299
18. Khuri RN, Agulian SK, Boulpaep EL, Simon W, Giebisch GH (1978) Changes in the intracellular electrochemical potentials of $Na^+$, $K^+$ and $Cl^-$ in single cells of the proximal tubules of the Necturus kidney induced by rapid changes in the extracellular perfusion fluids. Drug Res 28:879

19. Khuri RN, Agulian SK, Giebisch G (submitted) Electrochemical potentials
    of $Na^+$ and $K^+$ in the distal tubule of Amphiuma. Am J Physiol
20. Khuri RN, Agulian SK, Kalloghlian A (1972) Intracellular potassium in
    cells of the distal tubule. Pfluegers Arch 335:297
21. Khuri RN, Agulian SK, Nassar R (submitted) Electrochemical potentials
    of chloride in single fibers of mammalian skeletal muscle in vivo.
    Pfluegers Arch
22. Khuri RN, Bogharian KK, Agulian SK (1974) Intracellular bicarbonate in
    single skeletal muscle fibers. Pfluegers Arch 349:285-294
23. Khuri RN, Hajjar JJ, Agulian SK (1972) Measurement of intracellular
    potassium with liquid ion-exchange microelectrodes. J Appl Physiol
    32:419
24. Khuri RN, Hajjar JJ, Agulian SK, Bogharian K, Kalloghlian A, Aklanjian D,
    Bizri H (1972) Intracellular potassium in cells of the proximal
    tubule of Necturus. Pfluegers Arch 338:73
25. Khuri RN, Kalloghlian A, Agulian SK (submitted) Intracellular potassium
    in rat muscle under different metabolic states. Pfluegers Arch
26. Kimura T, Urakabe S, Yuasa S, Miki S, Takamitsu Y (1977) Potassium ac-
    tivity and plasma membrane potentials in epithelial cells of toad
    bladder. Am J Physiol 232:F196
27. Kostyuk PG, Sorokina ZA, Kholodova YuD (1969) Measurement of activity
    of hydrogen, potassium and sodium ions in striated muscle fibers and
    nerve cells. In: Lavallee M, Schanne OF, Hebert NC (eds) Glass Micro-
    electrodes. Wiley, New York pp 322-348
28. Lev AA (1964) Determination of activity coefficients of potassium and
    sodium ions in frog muscle fibers. Nature (London) 201:1132
29. Ling G, Gerard RW (1949) Measurement of the transmembrane electrical
    potential of frog sartorius muscle fibers. J Cell Comp Physiol 34:
    383-395
30. Rector FC, jr, Carter NW, Seldin DW (1965) The mechanism of bicarbonate
    reabsorption in the proximal and distal tubules of the kidney.
    J Clin Invest 44:278-290
31. Relman AS, Gorham GW, Levinsky NG (1961) The relation between external
    potassium concentration and the electrolyte content of isolated rat
    muscle in the steady state. J Clin Invest 40:386
32. Spring KR, Kimura G (1978) Chloride reabsorption by renal proximal
    tubules of Necturus. J Membr Biol 38:233-254
33. Steiner RA, Oehme M, Ammann D, Simon W (1979) Neutral carrier sodium
    ion-selective microelectrode for intracellular studies. Anal Chem
    51:351-353
34. White JE (1976) Intracellular potassium activities in Amphiuma small
    intestine. Am J. Physiol 231:1214

# Intracellular Ion Activity ($K^+$, $Ca^{2+}$ and $Cl^-$) and Membrane Potential of Frog Muscle in Vitro

M. DELPIANO, H. ACKER

## Introduction

Dufau et al. (3) decribed a double-barrel ion-sensitive microelectrode
(ISME) with an extra thin tip diameter (0.1 /um) developed for the simultaneous measurement of intracellular ion activity and membrane potential.
As the measurements performed with these electrodes had yielded extremely
low membrane potentials in cells in tissue culture (1), we decided to test
the accuracy of the electrodes with measurements of intracellular $K^+$,
$Ca^{2+}$ and $Cl^-$ activities of frog muscle, where the membrane potential
and ion concentration are known.

## Methods

The frog muscle (tibialis anterior longus) was prepared according to the
method of Ling and Gerard (9).The muscle in vitro was placed in a plexiglass chamber and superfused with frog Ringer solution (NaCl 4.20; CaCl
0.23; $MgSo_4$ 0.15; $NaHCO_3$ 3.21 g/litre) at $24^{\circ}C$, and equilibrated with
95% $O_2$ and 5% $CO_2$ in a separate vessel connected to the superfusion
chamber by a glass tube.

Intracellular activities for $Ca^{2+}$, $K^+$, and $Cl^-$ were measured with double-
barrel ion-sensitive electrodes (Theta-Glas, Kugelstatter, Garching).
One barrel filled with 1 M magnesium acetate was used to record the membrane potential. Mg acetate was chosen to depress the ion sensitivity of
the thin glass tip. The other barrel was filled with ion exchanger for
either calcium (Stern Comp.), potassium (Corning comp.) or chloride (Orion
Comp.). The electrode was calibrated before and after impaling a set of at
least 10 muscle fibers: $K^+$-ISME as well as $Cl^-$-ISME with solutions of
10 - 100 mM KCL, and $Ca^{2+}$ - ISME with $CaCl_2$ at concentrations between 2
and 0.2 mM (3). For the measurement of both ion activity and membrane
potential the electrical circuit described by Lensing and Sasse (8) was
used.

## Results and Discussion

Fig. 1 shows an original recording of three measurements of $Ca^{2+}$, $K^+$
and $Cl^-$ activities and the respective membrane potentials. It has been
found that the ion response was relatively quick, and that the values for
the membrane potentials were lower than those stated in the literature. To
test whether these differences were produced by the ion-sensitive channel,
at the end of each experiment electrodes were used without the ion exchanger, filled only with 3 M KCl or 1 M Mg-acetate.

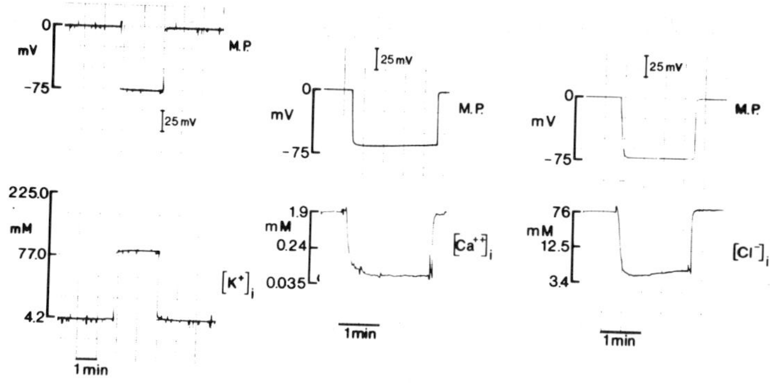



Fig. 1. Original recording of membrane potential (upper panel) and intracellular activities ([ ] i) of  $K^+$ , $Ca^{2+}$  and $Cl^-$  (lower panel) recorded in the frog muscle (tibialis anterior longus). Variation of membrane potential is given in mV changes from the isoelectric line; variation of ion activity in mM

Fig. 2 shows that the membrane potentials measured with the electrodes filled with 3 M KCL, which we used as a standard, varied from fiber to fiber between 67 mV and 105 mV. The mean value of 85.5±9.2 mV agrees with the values reported in the literature (4; 11). The electrodes filled with 1 M Mg-acetate alone measured membrane potentials that had a mean 3 mV lower than the values recorded with the 3 M KCl electrodes (mean 82.5+5.4 mV; range 70-98 mV) but not significantly different from them at the 98% level. Additionally, at the 98% level, Fig. 2 shows that each of the ion-selective electrodes applied for measurements of either $K^+$, $Ca^{2+}$, or $Cl^-$ activity recorded the membrane potential significantly lower (8 to 19 mV) than the electrodes without ion-exchanger. These results suggest that the respective ion exchanger influences the membrane potential of the muscle cells in a way so far we cannot explain. Furthermore, the experiments show that for $K^+$ and $Ca^{2+}$ the values for the membrane potentials are lower than those recorded with $Cl^-$-ISME. The question of whether the ion exchanger produces an ionic shift within the cell which perhaps depolarizes the cell must be investigated. The electrolyte applied for the measurement of the membrane potential cannot be responsible for these changes, because the electrodes filled with 1 M Mg-acetate alone show significantly higher values. In this connection it should be mentioned that the $Cl^-$-ISME records higher values for the membrane potential than do $Ca^{2+}$ and $K^+$-ISME. On the other hand, it seems that Cl ions do not appreciably influence membrane potentials in the frog muscle (10). Another factor interfering with this measurement is possibly the fact that the ion-sensitive channel has its own potential which intracellulary may change further. Such an offset-potential may influence the membrane potential as it is known

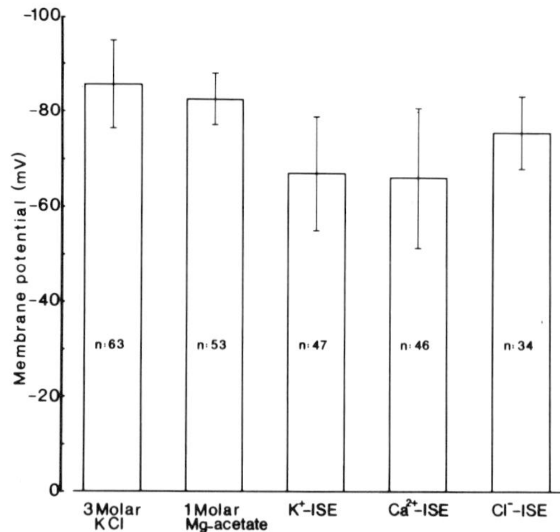



Fig. 2.  Mean values with S.D. of membrane potentials measured a) with ISE and b) with electrodes filled with 3 M KCl or 1 M Mg acetate.
Ordinate: membrane potential in mV, abscissa: different electrode types

from KCl electrodes (2; 5).

The intracellular $K^+$ activities range from 41 mM to 147 mM with a mean value of 110.05 ± 29 mM and show a linear relationship to the membrane potential but with different slopes (Fig. 3). The average value for intracellular $K^+$ activity agrees in its order of magnitude with the values reported by Khuri et al. (7), but is slightly higher than the values recorded by other authors (11). This discrepancy may be due to different qualities of the muscle preparations or to different functional states.

The intracellular $Ca^{2+}$ show considerable variation, ranging from 0.0012 mM to 0.096 mM with a mean value of 0.039±0.02 mM. This could mean that calcium ions originate from different compartments of the muscle cell or that the variations are produced by damage to the cell, or perhaps by other ions influencing the calcium response.

The $Cl^-$ activities range from 1.1 mM to 10.2 mM with a mean value of 3.51 ± 2.07 mM. The mean value is in good agreement with the values reported by Kernan et al. (6).

Summarizing, we were able to show that the ion-sensitive microelectrodes (tip diameter 0.1/um) measured the intracellular $K^+$ and $Cl^-$ activities in the range reported in the literature, whereas the values for $Ca^{2+}$ scattered between $10^{-6}$ and $10^{-4}$ M. It seems that the ion exchanger itself exerts an influence on the membrane potential.

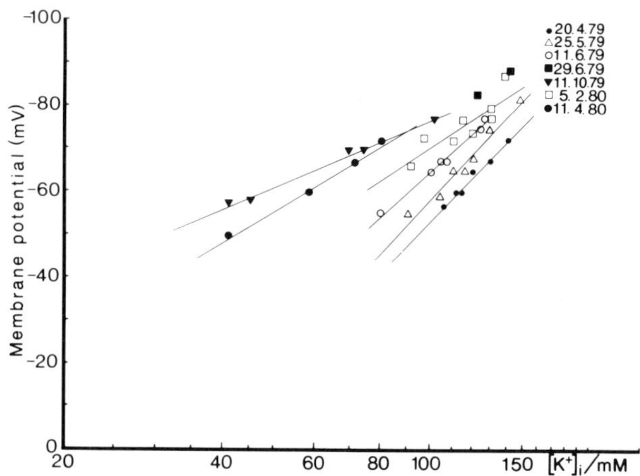



Fig. 3. Membrane potentials plotted against intracellular potassium activity: Each point represents the average data of three to five determinations performed in the same muscle. The regression lines connect the values
obtained from different points of the same muscles studied in an experiment.
Ordinate: membrane potential in mV, abscissa: intracellular $K^+$ activity
($[K^+]$) in mM (logarithmic scale)


## References


1. Acker H, Pietruschka F (1978) Membrane potential of cultivated type-I
     cells and its relationship to intracellular potassium and calcium.
     Pfluegers Arch 377:R 20
2. Adrian RH (1956) The effect of internal and external potassium concentra-
     tions on the membrane potential of frog muscle. J Physiol (Lond)
     133:631-658
3. Dufau E, Acker H, Sylvester D (1980) Double-barrel ion-sensitive micro-
     electrodes with extra thin tip diameters for intracellular measure-
     ments. Med Progr Technol 7:35-39
4. Haas J (1963) Die Physiologie der Muskelzelle. pp  197-244, Bornträger,
     Berlin.
5. Hironaka T, Morimoto S (1979) The resting membrane potential of frog
     Sartorius Muscle. J Physiol (Lond) 297:1-8
6. Kernan P, MacDermott M, Westphal W (1974) Measurement of chloride
     activity within frog sartorius muscle fibres by means of chloride-
     sensitive micro-electrodes. J Physiol (Lond) 241:60P-61P
7. Khuri R, Hajjar J, Agulian K (1972) Measurement of intracellular
     potassium with liquid ion-exchange microelectrodes. J Appl Physiol
     32:419-422

8. Lensing I, Sasse L (1978) Gleichzeitiges Messen des Sauerstoffpartial-
    drucks und bioelektrischer Potentiale im lebenden Gewebe. Elektronik
    27: 91-94
9  Ling G, Gerard W (1949) The normal membrane potential of frog sartorius
    fibers. J Cell comp Physiol 34:383-396
10. Stefani E, Steinbach B (1969) Resting potential and electrical proper-
    ties of frog slow muscle fibres. Effect of different external
    solutions. J Physiol (Lond) 203:383-401
11. Walker L, Brown M (1977) Intracellular ionic activity measurements
    in nerve and muscle. Physiol Rev 57:729-774

Max-Planck-Institut für Systemphysiologie, Rheinlanddamm 201, 4600 Dortmund
1 /FRG

# Adenosine 3'-5' Cyclic Monophosphate (cAMP)-Induced Changes in Intracellular Ionic Activities: Relation to Epithelial Ion Transport

W.McD. ARMSTRONG, A. DIEZ DE LOS RIOS, N.E. DeROSE

Ion-selective microelectrodes have been used successfully to measure steady-state intracellular ion activities in a variety of cells. In recent years, the development of liquid ion-exchanger microelectrodes that are easily constructed, have rapid response times, and are highly selective for individual ions of physiological importance, e.g. $K^+$, $Na^+$, $Ca^{2+}$, $Cl^-$, has opened up the possibility of investigating changes in intracellular activity that accompany or may be responsible for such cellular events as: excitation, contraction, secretion, and the action of specific hormones. This report describes a study currently under way in our laboratory in which the measurement of intracellular ionic activities is being used to obtain insights into the mechanism or mechanisms by which adenosine 3'-5' cyclic monophosphate (c-AMP) regulates ionic transfer processes in the small intestine.

## c-AMP and Epithelial Function

Many of the major organs of the body (e.g. the kidney, intestine, gallbladder, lungs, pancreas, salivary glands etc.) are lined by layers of epithelial cells. In a broad sense these epithelial layers serve two functions. One of these is to separate two solutions of widely different composition. More importantly, the difference in composition between these two solutions (e.g. blood and tubular fluid in the kidney) is generated and maintained by a complex array of secretory and absorptive transport processes within the epithelial layers themselves. Thus, transepithelial transport of water, ions, and other solutes plays a crucial role in homeostasis. The role of factors that modify or regulate transepithelial transport is of equal importance in this respect. These factors include external, neural and hormonal stimuli. In many instances (e.g. neural stimulation of salivary gland secretion, stimulation of water reabsorption in the kidney by anti-diuretic hormone) the external signal is received at one surface of the epithelial cell (the basolateral surface) and serves to activate or modulate processes located at the opposite (apical) surface. The concept of second messengers, i.e. intracellular agents that serve to transmit information from one cell surface to the other has therefore become of prime importance in the physiology of epithelial secretion. $Ca^{2+}$ and the cyclic nucleotides, c-AMP and c-GMP, appear to play a central role in this respect (2).

In mammalian intestine, c-AMP, or agents that increase cellular c-AMP levels (theophilline, prostaglandins, bacterial entertoxins, vasoactive intestinal peptide) induce a marked secretion of fluid and electrolytes (6, 16). The action of c-AMP on the intestine appears to consist of two different effects. One of these is an antiabsorptive effect: the inhibition of coupled electroneutral Na-Cl transfer from the luminal fluid across the apical membrane of the absorptive cells. The other is a direct stimulation of anion secretion. It has been suggested (7) that these two effects of c-AMP occur in different cell types of the intestinal epithelium the anti-

212

absorptive effect in the villous cell and the secretory process in the crypt cell. This suggestion is supported  by the finding that in flounder intestine (8) and in bullfrog small intestine (1) in which true functional crypts appear to be lacking, only the anti-absorptive effect of c-AMP is seen. On the other hand, anion secretion in response to c-AMP has been found in the intestine of the urodele amphibian Amphiuma in which the presence of functional crypts has not been established (15). Thus, the question of the specific cellular origin of the secretory and anti-absorptive effects of c-AMP on intestinal ion transport remains open.

A second question concerns the role of $Ca^{2+}$ as a mediator of c-AMP induced changes in intestinal transport. Addition of the $Ca^{2+}$ ionophore A 23187 to the solution bathing the serosal surface of stripped isolated preparations of rabbit ileum (3) or to the mucosal side of the isolated rabbit colon (9) produced changes in transepithelial ionic fluxes that were qualitatively similar to those elicited in the same tissue by c-AMP, although the ionophore was effective only when $Ca^{2+}$ was present in the external medium. These results strongly suggest that the action of c-AMP on ionic fluxes in intestinal epithelia is mediated by an increase in cytoplasmic $Ca^{2+}$. However, the question of whether the anti-absorptive effect of c-AMP, its secretory effect, or both of these are mediated by $Ca^{2+}$ again remains open. The experiments reported herein are the results we have obtained so far in an effort to solve these questions. Our approach to the problem was based on the following considerations. First, the ability of microelectrode techniques to monitor electrical and ionic activity changes in a single cell permits one, providing that the cell type in which these measurements are made can be positively identified, to obtain unequivocal data concerning the cellular responses to c-AMP. Second, the use of gallbladder epithelium which has only one type of cell, normally displays an absorptive function only, and in which the anti-absorptive effect of c-AMP alone has been identified (10) allows one to examine this process in a situation where it is not complicated by direct secretory responses.


## Effect of c-AMP and of A 23187 on Ionic Activities in Gallbladder

Isolated gallbladders from Necturus maculosus were used in this study. The animal was killed by a blow on the head. The gallbladder was quickly excised, opened, rinsed free of bile and mounted as a flat sheet at $23^{o}C$ in a divided chamber. The chamber was a modification of that described by White and Armstrong (26). The mucosal and serosal sides of the tissue were perfused continuously and independently. As described elsewhere (13), the perfusion system permitted the solutions bathing the mucosal and serosal aspects of the tissue to be changed rapidly. The reservoirs containing these solutions were bubbled with 100% $O_2$ throughout the experiment. The bathing solution contained, in mM, NaCl 100, $KH_2PO_4$ 0.8, $K_2HPO_4$ 2.3, Ca gluconate 1.8, and mannitol 21. The pH was 7.2. Adenosine 3'-5' cyclic-monophosphoric acid, (c-AMP) was added, at a concentration of 6 mM, to the serosal bathing medium. When this was done, the pH was adjusted to 7.2 with Tris (trishydroxymethylamino-methane) and, to maintain the same osmolality as that of the mucosal medium, the mannitol content was reduced to 9mM. Stock solutions of A 23187 were prepared by dissolving the ionophore in 95% ethanol. The final concentration of ethanol in the bathing medium was 0.1%. It was established by separate experiments that ethanol, at this concentration, did not affect the electrophysiological characteristics of the tissue. The transepithelial potential difference ($E_{Tr}$) was measured throughout the experiment by two calomel half cells connected through 3M KCl/agar bridges to the mucosal and serosal bathing solutions. The mucosal half-cell was grounded and used as a reference electrode. Microelectrodes were connected through an Ag/AgCl wire and a guarded cable to a high impedance ( $10^{14}$ ) preamplifier. Transepithelial and microelectrode potentials were recorded simultaneously on digital voltmeters and a strip-chart recorder.

Single barrelled micropipettes were drawn to a tip diameter of about 1 $\mu$m from capillary glass tubing previously cleaned by boiling in a concentrated detergent solution. Micropipettes for fabricating open tip and liquid ion-exchanger microelectrodes were drawn under identical conditions. Open-tip micropipettes for measuring apical membrane potentials ($E_m$) were back-filled with 3M KCl. Their tip resistances were about 20 - 30 M$\Omega$ . Liquid ion-exchanger microelectrodes were used to measure intracellular $K^+$, $Na^+$ and $Cl^-$ activities ($a_K^i$, $a_{Na}^i$, $a_{Cl}^i$). Corning 477317 and 477315 liquid ion exchangers were used in $K^+$ and $Cl^-$-selective microelectrodes. $Na^+$-selective microelectrodes were as described elsewhere (17). Liquid ion-exchanger microelectrodes were prepared by the method of Fujimoto and Kubota (12) as modified by Garcia-Diaz and Armstrong (13). They were calibrated in electrolyte solutions with concentrations covering the physiological range. Details of the calibration procedure, determination of selectivities, and the measurement of intracellular ionic activities have been given elsewhere (13). In the present study, it was found by direct measurement that the addition of 6mM c-AMP to the calibrating solutions did not affect the response of $K^+$, $Na^+$, or Cl-selective microelectrodes. Individual experiments were performed as follows: After mounting the tissue under open circuit conditions in the chamber, 60 min were allowed for the establishment of a steady state under control conditions. $E_m$ and the intracellular activity of one ion ($K^+$, $Na^+$, or $Cl^-$) were then measured. Following this, one of the bathing solutions was changed to another that contained either c-AMP (serosal side only) or A 23187 (either the mucosal or serosal side), 45 - 60 min were allowed for the establishment of a new steady state, and the above parameters were again measured. $E_{Tr}$ was recorded continuously throughout the experiment. Microelectrodes were mounted perpendicularly to the tissue on a micromanipulator (MM33, Narishige, Japan). This was used to bring the microelectrode close to the mucosal surface of the tissue. Final movement of and cell impalement with the microelectrode was accomplished under microscopic observation (Stereomicroscope III, Zeiss, New York, N.Y.) with a hydraulic drive micromanipulator (MO-10, Narishige, Japan).

During experiments with open-tip microelectrodes a repititive 0.5 nA current pulse was applied. In this way the input resistance of the microelectrodes used to measure $E_m$ was continuously monitored. Criteria for accepting impalements with open-tip and ion-selective microelectrodes were as reported elsewhere (13).

The results obtained with c-AMP are summarized in Table 1. Under control conditions, the average value found for $E_m$ (52 mV, inside negative) was in excellent agreement with the value (52.6 mV) previously reported for the same preparation under identical conditions (13). Somewhat higher values for $E_m$ in _Necturus_ gallbladder were reported by others (11, 21, 25), but when the high $K^+$ permeability of the apical cell membrane (21) and the different external $K^+$ concentrations used are taken into account, these values are consistent with those reported in the present study.

Table 1.

Effect of cAMP

| | Em (mV) | $E_{Tr}$ (mV) | $a_{Cl}^i$ (mM) | $a_K^i$ (mM) | $a_{Na}^i$ (mM) |
|---|---|---|---|---|---|
| Control | $-52+2$ (13) | $-0.1+0.2$ (13) | $14+2$ (4) | $84+5$ (5) | $8.5+1.1$ (4) |
| + 6mM cAMP | $-50+2$ (13) | $+0.8+0.2$ (13) | $9+1$ (4) | $113+6$ (5) | $5.5+0.7$ (4) |
| P | n.s. | 0.01 | 0.05 | 0.01 | 0.05 |

Paired t-test has been used. Values are given as means $+$S.E.M. Number in parentheses is the number of tissues studied.

c-AMP had no effect on $E_m$ (Table 1). On the other hand, $E_{Tr}$ is increased from virtually zero to a small, serosal-positive value. Similar effects of c-AMP on $E_m$ and $E_{Tr}$ in isolated bullfrog small intestine were reported by Armstrong and Youmans (1).
The control values for $a_{Na}^i$ and $a_{Cl}^i$ shown in Table 1 are in good agreement with those previously reported from this laboratory (13). Under control conditions $a_{Cl}^i$ is above the value corresponding to electrochemical equilibration. This value ($a_{Cl}^{eq}$) is easily calculated from the relationship

$$E_m = (RT/F) \ln (a_{Cl}^{eq}/a_{Cl}^o)$$

where $a_{Cl}^o$ is the Cl activity in the bathing medium and R, T, and F have their usual meanings. For $E_m = -50$ mV and $a_{Cl}^o = 77$ mM, in the present experiments, one obtains $a_{Cl}^{eq} - 10$ mM. The ratio $a_{Cl}^i/a_{Cl}^{eq}$ found in the present experiments under control conditions ranged from 1.4 to 1.6. This is in good agreement with the value (1.8) previously reported under identical conditions (13). Reuss and his associates (22, 23) reported a value of 4.3 for this parameter. This discrepancy may be related to two factors, the absence of external $HCO_3^-$ in the present experiments (13) and the fact that Reuss and his co-workers used a higher pH value than that employed in the present study. It has recently been reported (24) that epithelial salt absorption increases as the pH of the bathing medium is increased.

In the presence of c-AMP, $a_{Na}^i$ and $a_{Cl}^i$ decreased (Table 1), $a_{Cl}^i$ to a value that is indistinguishable from that corresponding to transmembrane electrochemical equilibration. This is consistent with an inhibition, by c-AMP, of coupled transapical NaCl entry. This conclusion is further supported by the results obtained with $E_m$ and $E_{Tr}$. The absence of any measurable effect of c-AMP on $E_m$ indicates that the NaCl entry process inhibited by c-AMP is electroneutral. The small increase in $E_{Tr}$ is consistent with the abolition, due to inhibition of salt transfer of the small ( 1.2 mV, mucosal positive) diffusion potential across the intracellular junctional complexes observed in the actively transporting gallbladder (4). The origin of this diffusion potential may be related to the cation-selectivity of the paracellular shunt pathway and to the hypertonic NaCl solution in the intracellular spaces predicted by the standing osmotic gradient theory of Diamond and Bossert (5).

The mean $a_K^i$ and $a_{Na}^i$ values found in the present experiments (Table 1) agree with previous estimates of these parameters in <u>Necturus</u> gallbladder under control conditions (13, 14, 23). c-AMP significantly increases $a_K^i$ (Table 1). Two possible explanations for this are: (1) direct stimulation by c-AMP of the baso-lateral $Na^+$-$K^+$ pump in the cells, (2) inhibition, when the outwardly-directed transmembrane electrochemical driving force for Cl is abolished by c-AMP, of a coupled baso-lateral K-Cl efflux (19). Further experiments are needed to discriminate between these two possibilities.

In summary, it was found in these experiments that c-AMP lowers $a_{Na}^i$ and $a_K^i$ in epithelial cells of the gallbladder. In the presence of this agent, $a_{Cl}^i$ declined to a value ($a_{Cl}^{eq}$) identical to that calculated for passive distribution of this ion across the cell membrane. (It should be noted that, because of the very low value of $E_{Tr}$, $a_{Cl}^{eq}$ values for the apical and baso-lateral cell membranes respectively do not differ greatly). $a_{Na}^i$ decreased to a value close to that observed in Cl-free media containing 100 mM $Na^+$ (Garcia-Diaz and Armstrong, unpublished observations). There is now compelling evidence that, in gallbladder, $Cl^-$ entry, through the apical cell membrane, occurs virtually exclusively _via_ a coupled electroneutral symport with $Na^+$ and that the transapical $Na^+$ gradient is the driving force for intracellular $Cl^-$ accumulation (13). Thus the present results point strongly to inhibition of this coupled entry process as the major effect of c-AMP on gallbladder absorptive cells.

To determine if the anti-absorptive effect of c-AMP in _Necturus_ gallbladder depends on an increase in intracellular $Ca^{2+}$, the effect of the $Ca^{2+}$ ionophore, A23187, on $E_m$ and $a_{Cl}^i$ was investigated. The rationale behind these experiments was this: If the effects of c-AMP on these parameters are indeed a consequence of elevated cytosolic $Ca^{2+}$ levels then, following addition of A23187 to the bathing medium, one might expect to find changes similar to those shown in Table 1. Since the reported effects of A23187 on ion transport in ideal mucosa were observed when the ionophore was added to the serosal bathing medium (3) whereas, with the colon, these effects were obtained when A23187 was added to the mucosal medium (9), in the present experiments, this agent was added to both media in turn. The results obtained are summarized in Table 2. It is evident that, in the isolated gallbladder, significant effects were observed only when A23187 was present in the mucosal medium. Under these conditions the apical cell membrane potential, $E_m$, was strongly hyperpolarized. Since $E_{Tr}$ did not change, the baso-lateral membrane potential must have been hyperpolarized to approximately the same extent. A similar effect of A23187 on both the apical and baso-lateral cell membranes in _Necturus_ gallbladder was reported recently by Reuss, Bello-Reuss and Grady (20). These authors suggested that this effect was due to an increase in the $K^+$ conductance of both cell membranes and that this increase in conductance is in turn mediated by an increase in cytoplasmic $Ca^{2+}$. The effect of mucosal A23187 on $a_{Cl}^i$ (Table 2) is consistent with this explanation. It can be seen from this table that, although the absolute value of $a_{Cl}^i$ decreased in the presence of mucosal A23187, the transmembrane driving force for Cl ($E_m$-$E_{Cl}$) was not abolished as it was in the presence of c-AMP (Table 1). On the contrary, A 23187 resulted in an approximately 2-fold increase in this gradient (i.e. from 11 to 20 mV). An increase in the $K^+$ conductance of both cell membranes would result in a corresponding increase in the conductive $K^+$ fluxes from the cell interior to the external medium. This could, in turn, cause a decrease in coupled neutral K-Cl efflux. This uncoupling of $K^+$ and $Cl^-$ fluxes from cell to medium could then result in the establishment of a higher $Cl^-$ electrochemical potential difference between the cell interior and the external medium than is found under control conditions. An alternative possibility is that $Ca^{2+}$ directly stimulates coupled electroneutral NaCl entry. In view of current evidence that $Ca^{2+}$ cells acts mainly as a secretagogue in many epithelial systems, this seems unlikely.

In conclusion, the present experiments strongly suggest that the anti-absorptive effect of c-AMP in epithelial cells of gallbladder and intestine is not mediated by an increase in cytoplasmic $Ca^{2+}$ and that increased levels of cytoplasmic c-AMP and of calcium may produce different effects in these cells. Our results also imply that the anti-absorptive and secretory effects of c-AMP in intestinal epithelia are mediated by different mechanisms. Experiments in which cytoplasmic $Ca^{2+}$ is being measured directly with $Ca^{2+}$-selective microelectrodes (18) are now being performed to obtain further information about the role of c-AMP in epithelial ion transport.

Table 2.

Effect of A 23187 (0.5/ug/ml) on intracellular chloride accumulation

| | N | $E_m$(mV) | $E_{Tr}$(mV) | $a_{Cl}^i$(mM) | $a_{Cl}^{eq}$(mM) | $a_{Cl}^i/a_{Cl}^{eq}$ |
|---|---|---|---|---|---|---|
| **Group 1** | | | | | | |
| Control | 5 | −46±2 | +0.1±0.4 | 19±1 | 12±2 | 1.6±0.2 |
| +A23187 (mucosal side) | 5 | −67±1 | +0.1±0.4 | 12±1 | 5.1±0.2 | 2.3±0.2 |
| P | | 0.01 | n.s. | 0.05 | 0.01 | 0.05 |
| **Group 2** | | | | | | |
| Control | 5 | −50±2 | −0.1±0.1 | 18±2 | 11±1 | 1.6±0.1 |
| +A23187 (serosal side) | 5 | −50±3 | −0.1±0.1 | 20±2 | 11±2 | 1.8±0.1 |
| P | | n.s. | n.s. | n.s. | n.s. | n.s. |

Paired t-test was used. Values are mean $\pm$ S.E.M. N is the number of tissue studied.

<u>References</u>


1. Armstrong WMcD, Youmans SJ (1980) The role of bicarbonate and adenosine 3', 5'-mono-phosphate (c-AMP) in chloride transport by epithelial cells of bullfrog small intestine. Ann New York Acad Sci 341: 139
2. Berridge MJ (1979) Relationship between calcium and the cyclic nucleotides in ion secretion. In: Binder HJ (ed) Mechanisms of intestinal secretion. Alan R Liss, New York. p 65.
3. Bolton JE, Field M (1977) Ca ionophore-stimulated ion secretion in rabbit ileal mucosa: relation to actions of cyclic 3', 5'-AMP and carbamylcholine. J Memb Biol 35: 159
4. Curci S, Frömter E (1979) Micropuncture of lateral intracellular spaces of Necturus gallbladder to determine space fluid $K^+$ concentration. Nature 278: 355
5. Diamond JM, Bossert WH (1967) Standing gradient osmotic flow. A mechanism for coupling water and solute transport in epithelia. J Gen Physiol 50: 2061
6. Field M (1978) Cholera toxin, adenylate cyclase and the process of active secretion in the small intestine: the pathogenesis of diarrhea in cholera. In: Andreoli TE, Hoffman JF, Fauestil DD (eds) The physiological basis for disorders of biomembranes. Vol 5. Plenum Press, New York. p 877
7. Field M (1979) Intracellular mediators of secretion in the small intestine. In: Binder HJ (ed) Mechanisms of intestinal secretion. AR Liss, New York. p 83
8. Field M, Smith PL (1975) Ion transport in the small intestine of the winter flounder. Pseudopleuronectes americanus. Bull Mount Desert Isl Biol Lab 15: 34
9. Frizzel RA (1977) Active chloride secretion by rabbit colon: calcium-dependent stimulation by ionophore A 23187. J Memb Biol 35: 175
10. Frizzell RA, Dugas MC, Schultz SG (1975) Sodium chloride transport by rabbit gallbladder. Direct evidence for a coupled NaCl influx process. J Gen Physiol 65: 769
11. Frömter E, Diamond J (1972) Route of the passive ion permeation in epithelia. Nature New Biol 235: 9
12. Fujimoto M, Kubota T (1976) Physiochemical properties of a liquid ion-exchanger microelectrode and its application to biological fluids. Jpn J Physiol 26: 631
13. Garcia-Diaz JF, Armstrong W MacD (1980) The steady-state relationship between sodium and chloride transmembrane electrochemical potential differences in Necturus gallbladder. J Memb Biol (in press)
14. Graf J, Giebisch G (1979) Intracellular sodium activity and sodium transport in Necturus gallbladder epithelium. J Memb Biol 47: 327
15. Gunter-Smith PJ, White JF (1979) Contribution of villus and intervillus epithelium to intestinal transmural potential difference and response to theophylline and sugar. Biochim et Biophys Acta 557: 425
16. Nellans HN, Frizzell RA, Schultz SG (1974) Brush-border processes and transepithelial Na and Cl transport by rabbit ileum. Am J Physiol 226: 1131
17. O'Doherty J, Garcia-Diaz JF, Armstrong W McD (1979) Sodium-selective liquid ion-exchanger microelectrodes for intracellular measurements. Science 203: 1349
18. O'Doherty J, Youmans SJ, Armstrong WMcD, Stark RJ (1980) Calcium regulation during stimulus secretion coupling: continuous measurement of intracellular calcium activities. Science 209: 510

19. Reuss L (1979) Electrical properties of the cellular transepithelial
    pathway in Necturus gallbladder. III Ionic permeability of the
    baso-lateral cell membrane. J Memb Biol 47: 239
20. Reuss L, Bello-Reuss E, Grady TP (1980) Cyanide increases $K^+$ conduc-
    tance in gallbladder epithelial cell membranes. Fed Proc 39: 1710
21. Reuss L, Finn AL (1975) Electrical properties of the cellular transepi-
    thelial pathway in Necturus gallbladder: I Circuit analysis and
    steady-state effects of mucosal solutions ionic substitutions. J
    Memb Biol 25: 115
22. Reuss L, Grady TP (1979) Effects of external sodium cell membrane
    potential on intracellular chloride activity in gallbladder epithe-
    lium. J Memb Biol 51: 15
23. Reuss L, Weinman SA (1979) Intracellular ionic activities and transmem-
    brane electrochemical potential differences in  gallbladder epithe-
    lium. J Memb Biol 49: 345
24. Smith PL, Orellana S, Field M (1980) Role of medium pH in the regula-
    tion of chloride transport in the intestine of the winter flounder.
    Pseudopleuronectes americanus. Bull Mount Desert Isl Biol Lab 19:
    24
25. Van Os CH, Slegers JFG (1975) The electrical potential profile of
    gallbladder epithelium. J Memb Biol 24: 341
26. White JF, Armstrong WMcD (1971) Effect of transported solutes on
    membrane potentials in bullfrog small intestine. Am J Physiol 221:
    194

## Acknowledgements

The research reported herein was supported by grants AM12715 and HL 23332
from the U.S. Public Health Service. A.D.R. was a fellow of the Ministerio
Universidades y Investigacion, Spain.

Indiana University, School of Medicine, Indianapolis, IN, USA 46223

# Continuous Recording of $K^+$ and $Ca^{2+}$ Pancreatic Juice of Conscious Rats Provided with an Extracorporeal Pancreatic Loop

F.J. HABERICH

## Introduction

During the last two years our interest has been especially involved in the development and improvement of methodical and technical problems concerning mainly two points:

1. An animal model for physiological experiments, permitting continuous measurements and recording of several parameters of bile and exocrine pancreatic secretion in conscious rats.

2. An improved method for continuous analysis of electrolytes in the streaming secretes using a new construction of ion-selective flow-through electrodes.

In the following paper I will report on our first steps into this new fascinating and virgin territory of investigation.

## Materials and Methods

### Surgical Procedure

In female Wistar rats fed with a standard diet (Ssniff: 56% Carbohydrates, 24% proteins, 3% fat) and weighing 250 - 300 g BW, the common bile duct is ligated by two PVC tubes (0.5 i.d.), one just below the hepatic hilum deriving bile and a second into the duodenal end of the bile duct allowing a separate derivation of pancreatic juice. Two additional catheters of the same size were also inserted into the animal duodenum. All catheters were brought out through the dorsal body wall and protected by adequate dressing. Connecting the tubes coming from liver and pancreas with the dudenal tubes, extracorporeal loops for bile and pancreatic juice are formed, allowing recirculation of both secretes into the duodenum. This recirculation is very important for a normal flow pattern of pancreatic juice, because there is a negative feed back control of exocrine pancreatic secretion due to the existence of pancreatic juice itself in the duodenum. Between the experiments the rats live in normal animal cages with free access to food and water. Only during the experiments are the animals kept in restraining cages, which produces a certain amount of immobilization. During the early post-operative period the serum amylase and lipase activities are significantly high but rapidly decline afterwards normalizing within 4 days. Normal values for flow-rate and protein secreted are: flow rate, 1.63 $\pm$ 0.63 ml/h·kg and 52.5 $\pm$ 17.6 mg/h·kg protein. For details of the surgical method and the recovery period after operation see (6, 7).

## Experimental Arrangement

During the experiments the loops are opened and the flowrate of the pancreatic juice is recorded by a drop counter. After the mixing of bile and pancreatic juice both secretes are recirculated using a peristaltic pump.

Between the animal and the drop counter a minaturized, newly constructed system of ion-selective flow-through-electrodes may be interposed one after the other. The particular advantages of this new system are:

1. Construction of all electrodes for ions, $Po_2$, $pCO_2$, as standardized modules with identical dimensions, allowing complete interchangeability.

2. The flow-through canal is integrated in the modules.

3. Several modules placed one after the other will form a flat "rifle-bore" canal with 1 mm of diameter presenting good conditions for an undisturbed laminar flow.

4. Small volume requirements: about 5 $\mu$l per module.

This arrangement allows the continuous and simultaneous measurement and recording of flow rates and several ionic activities in bile and pancreatic juice in the conscious rat without any loss of secrete.

## Evaluation

In order to evaluate the recorded curves, it must be considered that flow is always recorded immediately as the pancreatic juice being secreted, but it needs a certain time to reach the first ion-selective electrode. This time-lag between the curves of flow and ion activities unfortunately is not constant. It depends on the dead spaces of the system: from pancreas to the first electrode, from the 1st electrode to the 2nd a.s.o. - and the actual flow rate. A point to point correction of the curves is necessary in order to achieve true synchronous data.

## Results

One single experiment as an example may serve to present the described method:

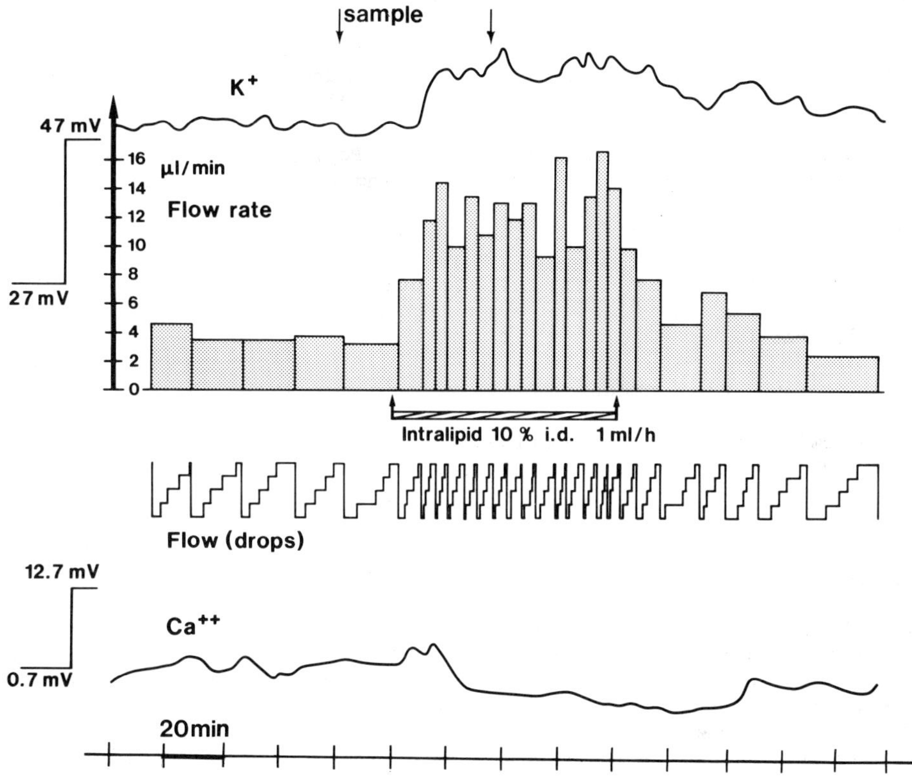



Fig .1.   For description and explanation of the figure see text

Fig. 1 shows original recordings of (from top to bottom): $K^+$-activity, calculated flow rate, drop counter curve (resetting after every 5 drops), and of $Ca^{2+}$-activity. Left of the activity curves are the calibrations. The evaluation of the curves includes several steps:

1. Calculation of flow rate (5 drops correspond to 65/ul of volume).

2. Correction of the activity curves with respect to time-lag due to the dead spaces from pancreas to the first electrode ($K^+$) ≙ 65/ul ≙ 5 drops and the second electrode ($Ca^{2+}$) about 75 - 80/ul ≙ 6 drops.

3. Calculation of mean concentration of $K^+$ and $Ca^{2+}$ /mM/1) for each flow period (5 drops) from the mean electrochemical potential (mV) recorded, according to the equation:

$$E(mV) = (\text{slope of electrode}) \cdot \log [K^+]$$

Therefore

$$[K^+] = 10^{(E/slope)}$$

The slope of each electrode must be individually determined. It is equivalent to the mV per decade of concentration.

The questions under investigation in these experiments were: Is there any relationship between volume flow rate and electrolytes secreted and are the electrolytes - esp. $Ca^{+2}$ - related to protein secretion. For this purpose we infused Intralipid as a physiological stimulant for exocrine pancreatic secretion - the intraduodenal (i.d.) infusion period is indicated by arrows in Fig. 1. Furthermore we took two samples (10/ul) for a protein determination, using a colorimetric biuret-method.

Flow Rate and $K^+$

It is very surprising to find that - contrary to the generally expressed opinion in the literature (2, 8) - there is a very pronounced linear relation between the amount of $K^+$ secreted per minute (nM/min) and the flow rate (/ul/min). The coefficient of regression is close to 1.0 (r = 0.98) (Fig. 2, upper). Since the amount of $K^+$ secreted is equal to the product of flow rate and concentration, one would expect that the relation between flow rate and $K^+$-concentration should be a hyperbolic function, if the other is strictly linear. As can be seen in Fig. 2 this is clearly not found.

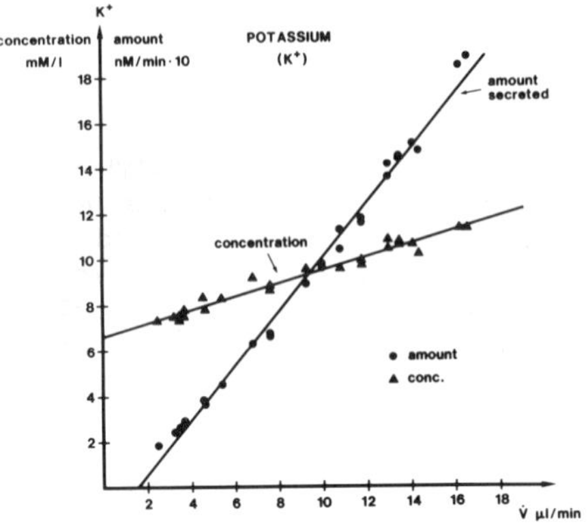


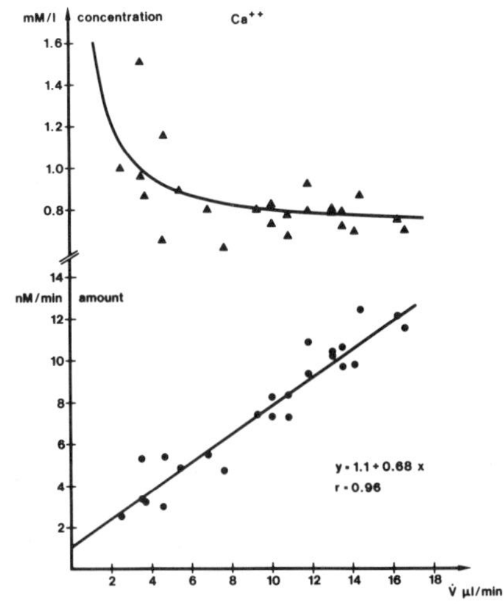


Fig. 2. For description and explanation of the figure see text

Flow Rate and $Ca^{2+}$

If we look at the lower part of Fig. 2 we find the corresponding relations for flow rate and $Ca^{2+}$-concentration as well as the amount of secreted $Ca^{2+}$ (nM/min). Once more, the amount is fairly well approximated as a linear function of flow rate (r = 0.977) and this time the expected hyperbolic function of the $Ca^{2+}$-concentration can be fairly clearly established, but indeed with a negative slope.

## Correlations

In coming back to the initial questions, it is very interesting and informative at the same time, to calculate some correlations between flow rate and the secretion of electrolytes and protein. During the experiment illustrated in Fig. 1 two samples were taken to determine protein, one before the infusion of intralipid and the other during the infusion. There is a high increase of flow rate as well as $K^+$ concentration, $K^+$ elimination and the amount of $Ca^{2+}$ secreted. Table 1 contains a list of correlated parameters, corresponding to the times of sampling. The first section of this table, the flow rate, makes clear that the factor, by which wolume flow is increased is the same as for the amount of protein secreted (/ug/min) and the amount of $Ca^{2+}$ secreted (nM/min). $K^+$ will not fit. The next section shows that there is no conformity in relation to the concentrations. We come to constant values, independent of flow rate of volume, protein and $Ca^{2+}$ amount if we calculate the amount of volume necessary to secrete 1/ug of protein or 1 mM of $Ca^{2+}$. As must be expected, the amount of $Ca^{2+}$ (nM) necessary to secrete 1/ug of protein is also constant, about 70. Calcula-

| Parameters | | | Control | Stimulated | Relation |
|---|---|---|---|---|---|
| Flow Rate (per min) | Fluid | µL | 3.6 | 12.0 | 1/3.33 |
| | Protein | µG | 207.6 | 685.3 | 1/3.30 |
| | $Ca^{++}$ | NM | 2.9 | 9.8 | 1/3.37 |
| Concentrations | $K^+$ | MM/L | 7.5 | 10.0 | 1/1.33 |
| | $Ca^{++}$ | MM/L | 2.12 | 0.78 | 2.72/1 |
| | Protein | MG/ML | 82.3 | 60.4 | 1.36/1 |
| Equivalents of Fluid | Protein | NL/µG | 16.8 | 17.5 | 1/1 |
| | $Ca^{++}$ | NL/NM | 1241 | 1224 | 1/1 |
| | $K^+$ | NL/NM | 144 | 96.7 | 1.5/1 |
| | $Ca^{++}$ | NM/NM | 68.9 | 68 | 1/1 |
| | $K^+$ | NM/NM | 8.0 | 5.4 | 1.5/1 |
| Protein versus | $Ca^{++}$ | µG/NM | 71.6 | 70 | 1/1 |
| | $K^+$ | MG/NM | 3.05 | 0.48 | 6.3/1 |

Table 1. For description and explanation of the table see text.

tion of the isomolar amounts come to the result that about 70 molecules of water are necessary for the secretion of 1 molecule of $Ca^{2+}$. For $K^+$ the number of water molecules under basal flow conditions is 8 while during increased flow this number is reduced to 5. If we believe that $K^+$ is mainly passively secreted, this reduction could be due to solvent drag under an increased volume-flow.


## Discussion

The surgical technique used is an improved modification of Colwell (4), Grossmann (5) and Pedrazolli (9). For details of surgery, experimental arrangement and recovery after operation see Haberich (6, 7).

In the ionselective flow-through-systems generally used (10), the tip of the electrodes protrudes into the canal perfused, which may cause disturbancies of flow and instabilities of the measured potential from the electrode due to streaming potentials.

As already mentioned before, it is generally assumed that the $K^+$-concentration of pancreatic juice is fairly constant and not dependent on the rate of volume flow (2, 8). Only Sewell and Young (11) have reported on an increase of $K^+$-concentration in the anesthetized rat and a compensatory fall of $Na^+$ during stimulation of pancreatic secretion with secretin. Our data would confirm this, although we already know that $K^+$-concentration is remarkably decreased by anesthesia alone.

The relation of $Ca^{2+}$-concentration and protein concentration was clearly established by Cecarelli et al. (3) and Argent et al. (1). Our findings confirm a better correlation between the amounts of $Ca^{2+}$ and protein secreted. Furthermore they permit us to formulate some stoichiometric relations between isomolar amounts of $Ca^{2+}$, fluid volume and protein secreted:

$$70 \text{ M } H_2O \triangleq 1 \text{ M } Ca^{2+} \triangleq 1/ug \text{ protein}$$

Furthermore we think that it seems more reasonable to establish relations between amounts and not between concentrations.

For a protein with a MW of 70 000, 1/ug would correspond to 1/uM, ergo: 1 Mol $Ca^{2+}$ for 1 Mol protein. It is not possible to calculate reasonable similar equivalents for isomolar $K^+$-amounts.


## References


1. Argent BE, Case RM, Scratcherd T (1975) Amylase secretion by the perfused cat pancreas in relation to the secretion of calcium and other electrolytes and as influenced by the external ionic environment. J Physiol 230: 575
2. Case RM, Harper AA, Scratcherd T (1969) Water and electrolyte secretion by the pancreas. In: Botelho SY, Brooks FP, Shelley WB (eds) The exocrine gland. Pennsylvania, University Press. pp 39-56
3. Cecarelli B, Clemente F, Meldolesi J (1975) Secretion of calcium in pancreas juice. J Physiol 245: 617

4. Colwell AR (1951) Collection of pancreatic juice from rats and consequences of its continued loss. Am J Physiol 164: 812
5. Grossmann MI (1958) Pancreatic secretion in the rat. Am J Physiol 194: 535
6. Haberich FJ, Bozkurt T, Reschke W (1980) Physiological studies of exocrine pancreatic secretion in conscious rats. 1. Communication: Surgical methods and registration techniques. (to be published in Gastro 8/80)
7. Haberich FJ, Bozkurt T (1980) Physiological studies of exocrine pancreatic secretion in conscious rats. 2. Communication: Recovery of the animal after operation. (to be published in Gastro 9/80)
8. Harper AA, Scratcherd T (1979) Physiology. In: Howatt HT, Sarles H (eds) The exocrine pancreas. WB Saunders Co, London, Philadelphia, Toronto. pp 50-85
9. Pedrazolli S, Dodi G, Varotto S, Lise M (1975) Technique for collection of pancreatic juice from rats. Rendic Gastroenterol 7: 31
10. Schindler JG (1977) Multimeßsystem für die elektrochemische Analyse strömender Flüssigkeiten und Gase. Biomed Techn 22: 235
11. Sewell WA, Young JA (1975) Secretion of electrolytes by the pancreas of the anesthetized rat. J Physiol 252: 379


## Acknowledgements

I would like to thank T. Bozkurt (laboratory assistant), Dr. B. Lingelbach (physics), A. Mazzola (drawings and photography), W. Reschke (animal surgery) and W. Riemann (material engineering and workshop) for their assistence and collaboration.

I would like to thank Prof. Dr. W. Simon from the ETH (Eidgenössische Technische Hochschule) Zürich, for providing us with the selective ion-exchange material for the construction of our electrodes.


Institut für angewandte Physiologie der Philipps-Universität Marburg/Lahn, Lahnberge, 3550 Marburg/Lahn

# Measurements of Potassium Activities and Membrane Potentials in Tumor Cells

H. ACKER, J. CARLSSON

It has been shown by Thomlinson and Gray (7) and Tannock (6) that only tumor cells near capillaries were proliferative and that cells at a distance of about 100 - 200 $\mu$m degenerate. These two studies were carried out on tumors with a pronounced nodular appaearance. Three-dimensional cell colonies, spheroids growing in suspension have been used in a model for nodular tumor growth (3, 9). With these models it might be possible to understand whether an efficient supply of oxygen and nutrition, a sufficient clearance of catabolic products, optimal tonicity, pH and ion strength can induce proliferation or a lack of these factors a degeneration. The spheroids contain an outer layer of mainly proliferative cells, an intermediate layer with low proliferative activity but with morphologically intact cells and, if the spheroids are big enough, a central necrotic region. We used different human tumor cells for our studies (3) (Osteosarcoma OS393, thyroidea carcinoma Hth 7, Glioma Mg 118) and for comparison embryonic hamster lung cells (V 79). Measurement of both intracellular potassium activity and membrane potentials of these cells, might provide a better understanding of the induction of cell death in poorly vascularized regions and a better knowledge of the metabolic state of the morphologically intact cells surrounding the necrotic zone. This knowledge is of importance since tumor cells in poorly vascularized regions show an increased resistance to radiation when they are deprived of oxygen (1) and they are difficult to reach with cytotoxic substances. These cells may therefore contribute to tumor growth during clinical treatment of patients.

For measuring the electrophysiological data the spheroids attached to a thin coverglass were mounted in a superfusion chamber. The spheroids were superfused with a culture medium without serum. The oxygen pressure in the medium of about 115 mm Hg, pH 7.4 and temperature 37$^{\circ}$C were continuously controlled. The electrode measurements were done in this perfusion chamber under microscopic control. The double-barrel electrodes as described by Dufau et al. (2) with tip diameters of about 0.1 $\mu$m were used to measure membrane potential and intracellular potassium activity simultaneously. The position of the electrode was electronically measured through a potentiometer attached to a manipulator system (David Kopf Instruments). Registration of potential and potassium in a glioma spheroid, are shown in Fig. 1. When the electrode first entered a peripheral cell in the spheroid stable potential and potassium signals were obtained. After one minute, the electrode was advanced about 30 $\mu$m and during the movement both potential and potassium signals returned to the levels for the incubation medium. At the end of the movement a new cell was hit. The signals obtained for potential and K$^+$ were not stable that time. The signal changed continuously toward background values indicating that the cell was ruptured or only touched upon. Thus only a peak value was obtained. Thereafter the electrode again advanced into the spheroid. After about 10 $\mu$m a new cell was hit and stable signals for potential and K$^+$ were obtained. During further penetration a peak value was obtained and then the electrode gave background incubation medium readings indicating that the necrotic zone was reached. Both

stable and peak value were evaluated. This was necessary to allow an esti-
mation of the relative number of electrophysiologically active cells at
different depths in the spheroids.

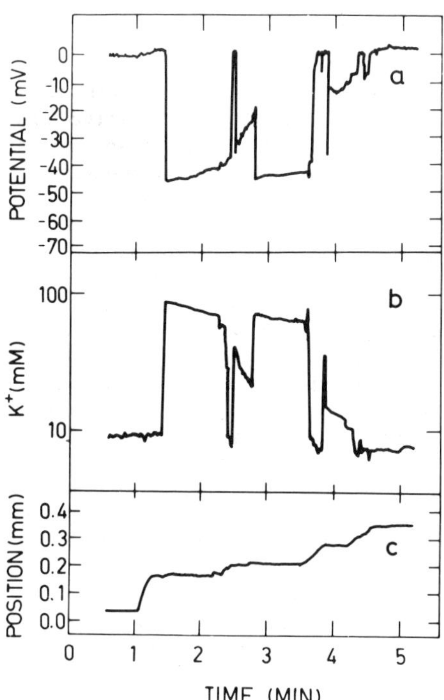


Fig. 1. Simultaneous recording of membrane potential and potassium activity in a glioma spheroid. At the bottom the position of the electrode is
electrically monitored

Table 1.

|  | MP/mV | $K^+$/mM |
|---|---|---|
| glioma MG 118 | − 42 + 6 | 80 + 26 |
| thyroidea Hth 7 | − 35 + 6 | 80 + 25 |
| osteosarcoma Os 393 | − 28 + 6 | 70 + 25 |
| V 79 | − 24 + 6 | 87 + 27 |

In Table 1 the mean values for membrane potentials and intracellular
potassium activity of the four different spheroids are shown. The absolute
values of the stable potentials were somewhat low. Normal glia cells, for
example, have membrane potentials of about −75 to −90 mV and the resting
potential of neurons have been reported to be −50 to −75 mV, for cells of
different species (5). Hause et al. (4) measured in malignant and nonmalignant cells growing as monolayers values from −12 to −35 mV which is in the

same range as our values. Timmerman and Buttlar (8) found membrane potentials of -10 mV in human carcinoma cells directly after explantation from rectum. The intracellular potassium activities measured in the spheroids are in the range of the values reported from cardiac muscles (frog and rabbit) (10). The intracellular concentration of potassium for normal glia cells have been reported to be 100 to 110 mM (5). Assuming an intracellular activity coefficient of about 0.6 to 0.8 (10), the potassium activity will be about 60 to 90 mM which is in the same range as that recorded in our glioma cells. No potentials were found in the necrotic centre indicating a short circuit between the necrotic centre and the outside of the spheroids. The potassium activity was about the same in the necrotic centre as in the culture medium in all spheroids. That means that potassium originating from cells undergoing cell death near the necrotic centre can diffuse freely and can obtain equlibrium with the medium. Membrane potential and intracellular potassium activities have different relationships in different tumor cells, as shown in Figure 2. The rather close relation between potassium and potential for glioma cells indicates a direct dependance of both. This is in agreement with the fact that the membrane potential of normal glia cells is mainly determined by the potassium activity (5). The membrane potential of the other tumor cells might be also determined by other ions, which could be assumed by the different shapes of the correlation between membrane potential and intracellular potassium activity.

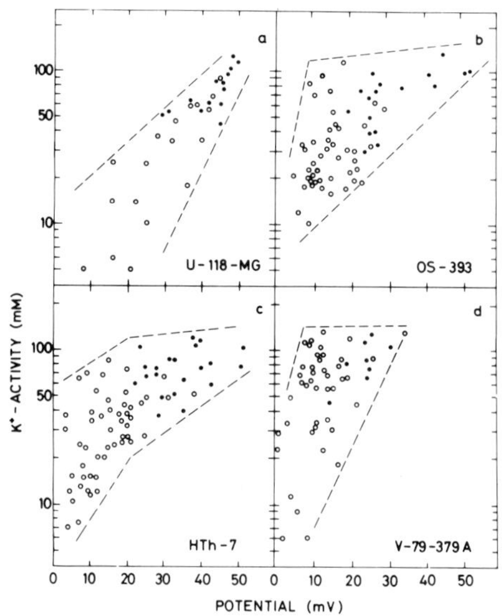


Fig. 2. Relationship between membrane potential and intracellular potassium activity in different tumor cells

The relative number of electrophysiologically active cells decreased drastically with depth in all spheroids, which can also be correlated with the relative number of proliferative cells. The large amount of morphologically inactive cells in the deeper regions must have been electrophysiologically inactive, probably indicating that only cells with electrical ac-

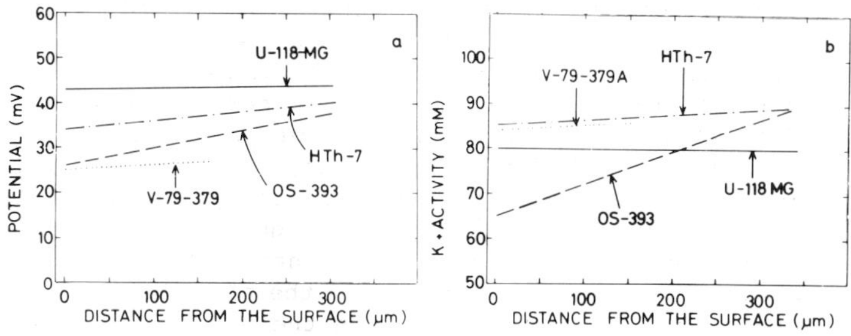

Fig. 3. Membrane potential and intracellular potassium activity as a function of depth in different spheroids


tivity can proliferate. No significant variation in the amplitude values of the potentials or the intracellular potassium activity could be detected as a function of depth except in the osteosarcoma which has an increase of both parameters with depths, as shown in Fig. 3. This indicates that electrophysiological death might be an all or nothing phenomenon. If the cells were active, then they would have normal values, otherwise no signal could be measured. These cells might not only exist in spheroids but also in poorly vascularized tumor tissue. It has been speculated that hypoxic and therefore radiation protected cells could be situated in the poorly vascularized regions (1). From our study it seems that the fraction of tumor cells in poorly vascularized regions having normal metabolism are much fewer than can be estimated just on a morphological basis. Thus, the amount of cells having the capacity to contribute to growth might be overestimated when inspecting histological sections.


## References


1. Denekamp J, Fowler RF (1977) Cell proliferation kinetics and radiation therapy. In: Becker FF (ed) Cancer, vol 6. Plenum Press, New York, p 105
2. Dufau E, Acker H, Sylvester D (1980) Double-barrel ion-sensitive microelectrodes with extra thin tip diameters for intracellular measurements. Med Progr Technol 7: 35
3. Haji-Karim M, Carlsson J (1978) Proliferation and viability in cellular spheroids of human origin. Cancer Research 38: 1457
4. Hause LL, Pattilo RA, Sances A, Mattingly RF (1970) Cell surface coatings and membrane potentials of malignant and nonmalignant cells. Science 169: 601

5. Kuffler SW, Nicholls JG (1976) From neuron to brain. A cellular approach to the function of the nervous system. Sinauer Press, Massachusetts
6. Tannock I (1968) The relation between cell proliferation and the vascular system in a transplanted mouse mammary tumor. Brit J Cancer 22: 258
7. Thomlinson RH, Gray LH (1955) The histologic structure of some human lung cancers and the possible implications for radiotherapy. Brit J Cancer 9: 539
8. Timmermann J, Buttlar M (1978) Membranpotentialuntersuchungen an menschlichen Epithelkarzinomzellen unter ionisierender Bestrahlung. Strahlentherapie 154: 700
9. Sutherland RM, Durand RE (1976) Radiation response of multicell spheroids. An in vitro tumor model. Current Topics in Rad Res, Quarterly vol 11, No : 87
10. Walker JL, Brown M (1977) Intracellular ionic activity measurements in nerve and muscle. Physiological Reviews 57: 729

Max-Planck-Institut für Systemphysiologie, Rheinlanddamm 201, 4600 Dortmund 1/FRG

# Ionized Calcium Secretion in the Duodenal Juice of Normals, Chronic Alcoholics and Patients with Different Stages of Chronic Alcoholic Pancreatitis


J. LOHSE, A. PFEIFFER, H. KAESS


## Introduction

The studies of Edmondson (1, 2) have established the pathogenetic role
of calcium in chronic alcoholic pancreatitis and pancreatic lithogenicity.
Additional investigations revealed increased concentrations of total (3,
4, 5, 6) and ionized (7) calcium in chronic calcified pancreatitis in comparison to healthy controls. In order to get to a further understanding
of pancreatic lithogenicity in chronic alcoholic pancreatitis, ionized
calcium and total calcium secretion were determined in the duodenal juice
of normals, chronic alcoholics without pancreatic desease and patients suf-
fering from chronic alcoholic pancreatitis of different stages: chronic
pancreatitis without injury of the exocrine function, non-calcified pancreatitis and calcified pancreatitis respectively, both presenting exocrine
insufficiency.

## Methods

Duodenal juice was continuously collected in 20-minute fractions for one
hour under simultaneous external stimulation with 1 U/kg/hr (U = units)
secretin and CCK respectively. Gastric juice was aspirated through a second
lumen to avoid endogenous pancreatic stimulation.

## Patients

20 healthy subjects served as controls (N) and none of them had a desease
of the gastrointestinal tract. 13 chronic alcoholics (A) had an alcohol
index of greater then 100 g/day lasting more than five years; all of them
suffered from alcoholic liver desease, but had no history or biochemical
signs of pancreatic affection. 33 patients had chronic alcoholic pancreatitis of different stages: 8 of them had a typical history of relapsing
hyperamylasaemia with accompanying painful attacks, but a normal pancreatic
function (Cp); 6 others had a similar history and presented insufficiency
of the secretin-dependant parameters volume and bicarbonate (NCPs); 10
showed furthermore reduced enzyme secretion ($NCP_{s+CCK}$). No calcifications
could be detected in the area of the pancreas by abdominal x-ray in all
these groups. The remaining 9 patients had radiologically proven calcifi-
cation of the pancreas and presented distinct exocrine insufficiency of all
parameters.

<u>Determinations</u>

Ionized calcium was measured by means of a calcium-selective macroelectrode (Philips, Netherlands), total calcium was determined by flamephotometry (Eppendorf, West Germany) and bicarbonate by the titrimetric method (Radiometer, Denmark). Amylase-, lipase-, trypsin- and chymotrypsin-activities were determined by standard photometric methods (Radiometer, Denmark).

<u>Data Analysis</u>

Values are given as means $\pm$ SEM. Student's t-test was used as appropriate for statistical analysis.

<u>Results</u>

Total calcium and ionized calcium concentrations are shown in Fig. 1. Chronic alcoholism leads to a significant increase of total ($p < 0.05$) and ionized calcium ($p < 0.02$) concentrations in the duodenal juice compared to healthy subjects. In the early stage of alcoholic pancreatitis (CP), calcium concentrations are not different from those of chronic alcoholics and controls. In later stages of the desease ionized calcium ($p < 0.001$) and total calcium ($p < 0.05$) concentrations increase progressively whereas pancreatic enzymes decrease to exocrine insufficiency (not shown); however, no statistical difference was observed between severe non-calcified and calcified pancreatitis in respect to their calcium concentration.

Similar results are obtained for ionized calcium and total calcium outputs (not shown) with the exception of calcium outputs in $NCP_s$-patients, which are inferior to those of CP-patients and chronic alcoholics and not different from controls, due to a reduced volume secretion. Ionized calcium outputs in the late stages of alcoholic pancreatitis ($NCO_{s+CCK}$ and CCP) were significantly ($p < 0.001$) enhanced compared to chronic alcoholics and early stages of chronic pancreatitis, although the total calcium outputs were not different from alcoholics, but significantly increased when compared to CP ($p < 0.003$)

During the secretin- and CCK-stimulation period the secretion of ionized calcium and total calcium (not shown) increased in all groups studied (Table 1). About 24% of total calcium was ionized in normals after one hour of exogenous hormonal stimulation, compared to a 12%-increase in severe alcoholic pancreatitis.

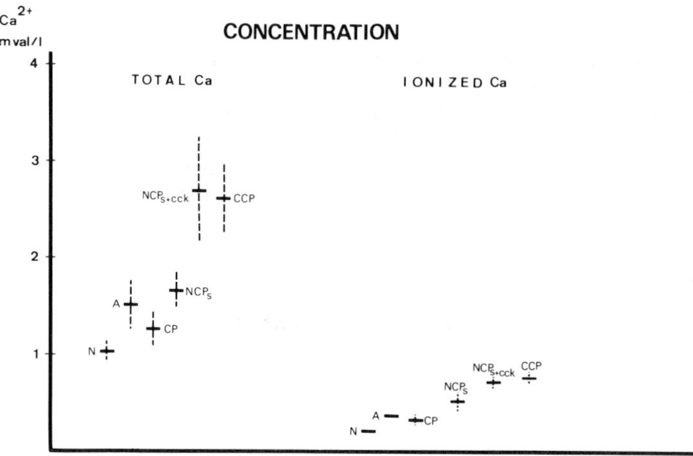



Fig. 1.  Total and ionized clacium concentrations in the duodenal juice of normals (N), chronic alcoholics (A), pancreatitis with a normal pancreatic function (CP) non-calcified pancreatitis (NCP$_S$ and NCP$_{S+CCK}$) and calcified pancreatitis (CCP) respectively

Ionized Ca in % of total Ca in duodenal juice

| patients | 20 min. | 40 min. | 60 min. |
|---|---|---|---|
| N | 15,6 | 20,6 | 23,9 |
| A | 23,7 | 23,5 | 26,9 |
| CP | 21,8 | 24,6 | 24,3 |
| NCP$_S$ | 30,6 | 29,9 | 34,7 |
| NCP$_{S+CCK}$ | 23,3 | 26,1 | 36,8 |
| CCP | 24,2 | 29,0 | 34,0 |

Table 1. Ionized calcium in % of total calcium in the duodenal juice

234

The percent changes of ionized calcium and total calcium secretion during
alcoholism and chronic alcoholic pancreatitis is seen in Table 2. The
ionized fractions increased more than the total calcium secretion, reveal-
ing highest rates in the first 20-min-fraction of CCP-patients.

% - increase of total and ionized Ca-concentration in duodenal juice

| patients | 20 min. | | 40 min. | | 60 min. | |
|---|---|---|---|---|---|---|
| | $Ca_T$ | $Ca^{++}$ | $Ca_T$ | $Ca^{++}$ | $Ca_T$ | $Ca^{++}$ |
| N | 100 | 100 | 100 | 100 | 100 | 100 |
| A | + 10 | + 67 | + 46 | + 67 | + 31 | + 47 |
| CP | + 10 | + 54 | + 23 | + 48 | + 56 | + 59 |
| NCP (s) | + 36 | +167 | + 61 | + 133 | + 75 | +153 |
| NCP s+CCK | +145 | +267 | +163 | + 233 | +130 | +253 |
| CCP | +212 | +383 | +154 | + 257 | +137 | +235 |

Table 2.  %-increase of total and ionized Ca-secretion in the duodenal
juice of chronic alcoholics and patients with different stages of chronic
alcoholic pancreatitis

## Discussion

Up to now it cannot be decided if the enhanced duodenal calcium secretion
with a predominant increase of the ionized fraction in chronic alcoholics
is due to liver-cell-damage or due to undetected pancreatic lesions,
since the secretin-CCK-test presents only pathological results when almost
70% of the pancreas is destroyed.

The progressive increase of duodenal calcium secretion in the course of
alcoholic pancreatitis, demonstrating a higher percentage increase of
the ionized fraction, may indicate its importance in pancreatic lithogenicity. This increase of ionized calcium may originate from the ionized
serum calcium, passing through lesions of the tight junctions between
the ductal cells into the pancreatic juice. Those lesions were demonstrated
by Nakamura (8) in chronic alcoholic pancreatitis in the insufficient
stage.

Another source of enhanced ionized calcium in the duodenal juice could
be a decrease of protein-bound calcium in exocrine insufficiency. Since
no linear relationship was found between pH and ionized calcium concentration, it is assumed that moderate pH-variations in the duodenal juice
do not affect ionized calcium concentrations.

The observation that no difference in calcium secretion was found between severe stages of non-calcified and calcified pancreatitis, suggests indirectly that additional factors, such as the pancreatic-stone-protein, which was extracted from human pancreatic stones (9,10), are also involved in pancreatic lithogenicity.


<u>References</u>


1. Edmondson HA, Bullock WK, Mehl JW (1949)  Chronic pancreatitis and
     lithiasis I. A clinicopathological study of 62 cases of chronic
     pancreatitis. Am J Pathol 25: 1227-1247
2. Edmondson HA, Bullock WK, Mehl JW (1950)  Chronic pancreatitis and
     lithiasis II. Pahology and pathogenesis of pancreatic lithiasis.
     Am J Pathol 26: 37-55
3. Goebell H, Bode Ch, Horn HD (1970)  Einfluß von Sekretin und Pan-
     creozymin auf die Calciumsekretion im menschlichen Duodenalsaft
     bei normaler und gestörter Pankreasfunktion. Klin Wschr 48: 1330
4. Goebell H, Horn HD, Grossmann HH, Bode Ch (1970) Calcium excretion
     in duodenal juice before and after application of secretin and
     pancreocymin in human. Proc III Sympos Europ Pancreatic Club Prague
     1968. Czechoslovak Medical Press Praha p. 174
5. Gullo L, Sarles H, De Barros Mott, Tiscornia O, et al. (1974)  Pancrea-
     tic secretion of calcium in healthy subjects and various diseases of
     the pancreas. Scand J Gastroenter 6: 35
6. Warwick R, Tothill P, Percy-Robb W (1973)  The calcium concentration in
     pancreatic secretion in chronic pancreatitis and carcinoma of the
     pancreas. Scand J Gastroent 8: 301
7. Lohse J. Hümmer H, Pfeiffer A, Kuntzen O, Kaess H (1979)  Ionized
     calcium and citrate secretion in the duodenal juice of normals
     and patients with chronic calcifying pancreatitis. 12th Meeting
     Europ Pancreatic Club Copenhagen 1979. Dan Med Bulletin 26: 35
8. Nakamura K, Sarles H, Payan H (1972)  Three-dimensional reconstruction
     of the pancreatic ducts in chronic pancreatitis. Gastroenter 62: 942
9. Lohse J, De Caro A, Sarles H (1979)  Partial characterization of a
     protein isolated from human pancreatic stones. 11th Meeting of the
     Europ Pancr Club 1978. Gastroenter Clin Biol 3: 308
10. De Caro A, Lohse J, Sarles H (1979)  Characterization of a protein
     isolated from pancreatic calculi of men, suffering from chronic
     calcifying pancreatitis. Biochem Biophys Res Comm 87: 1176-1182



5. Med. Abtlg., Städtisches Krankenhaus München-Schwabing, 8000 München/FRG

# Index

## Biochemistry of Membrane Transport

FEBS-Symposium No. 42
Editors: G. Semenza, E. Carafoli
1977. 392 figures, 76 tables. XIX, 669 pages
(Proceedings Life Sciences)
ISBN 3-540-08082-1

## Biochemistry

1979. 37 figures, 18 tables. IV, 193 pages
(Topics Current Chemistry, Volume 78)
ISBN 3-540-09218-8

## Biochemistry

1979. 84 figures, 20 tables. IV, 178 pages
(Topics Current Chemistry, Volume 83)
ISBN 3-540-09312-5

## Brain and Heart Infarct

Proceedings of the Third Cologne Symposium,
June 16–19, 1976
Editors: K. J. Zülch, W. Kaufmann, K.-A. Hossmann,
V. Hossmann
With contributions by numerous experts
1977. 155 figures, 14 tables. XVIII, 349 pages
ISBN 3-540-08270-0

## Brain and Heart Infarct 2

Editors: K. J. Zülch, W. Kaufmann, K.-A. Hossmann, V. Hossmann. With contributions by numerous experts
1979. 114 figures, 22 tables. XII, 330 pages
ISBN 3-540-09401-6

## Pathophysiology

Editors: A. A. Bühlmann, E. R. Froesch
With contributions by numerous experts
Translated from the German by T. C. Telger
1979. 74 figures, 84 tables. XII, 403 pages
(German Edition: Heidelberger Taschenbuch 101)
ISBN 3-540-90370-4

## Carbohydrate Metabolism in Pregnancy and the Newborn 1978

Editors: H. W. Sutherland, J. M. Stowers
1979. 95 figures, 177 tables. XIV, 558 pages
ISBN 3-540-08798-2

## Cerebral Circulation and Metabolism

Sixth International CBF Symposium, June 6–9, 1973
Editors: T. W. Langfitt, L. C. McHenry, Jr., M. Reivich,
H. Wollmann
1975. 180 figures, 100 tables. XXVIII, 566 pages
ISBN 3-540-06645-4
Distribution rights for Japan: Nankodo Co. Ltd., Tokyo

Springer-Verlag
Berlin
Heidelberg
New York

## Chemoreception in the Carotid Body

Editors: H. Acker, S. Fidone, D. Pallot, C. Eyzaguirre,
D. W. Lübbers, R. W. Torrance
1977. 101 figures, 21 tables. XIII, 296 pages
ISBN 3-540-08455-X

P. Deetjen, J. W. Boylan, K. Kramer
## Physiology of the Kidney and of Water Balance

Translator: R. V. Coxon
1975. 63 figures. IX, 141 pages
(Springer Study Edition)
ISBN 3-540-90048-9

## Effects of Ionizing Radiation on DNA

Physical, Chemical and Biological Aspects
Editors: A. J. Bertinchamps, J. Hüttermann, W. Köhnlein,
R. Téoule
With contributions by numerous experts
1978. 74 figures, 48 tables. XXII, 383 pages
(Molecular Biology, Biochemistry and Biophysics,
Volume 27)
ISBN 3-540-08542-4

## Fundamentals of Neurophysiology

Editor: R. F. Schmidt
With contributions by J. Dudel, W. Jänig, R. F. Schmidt,
M. Zimmermann .
Translated by M. A. Biedermann-Thorson from "Grundriss
der Neurophysiologie" 4th German edition
2nd revised and enlarged edition 1978. 137 figures.
IX, 339 pages
(Springer Study Edition)
ISBN 3-540-08188-7

## Fundamentals of Sensory Physiology

Editor: R. F. Schmidt
With contributions by numerous experts
Translated from the 3rd German Edition "Grundriss der
Sinnesphysiologie" by M. A. Biedermann-Thorson
1978. 138 figures, 5 tables. IX, 286 pages
(Springer Study Edition)
ISBN 3-540-08801-6

## Transport Mechanisms of Tryptophan in Blood Cells, Nerve-Cells, and at the Blood-Brain Barrier

Proceedings of the International Symposium,
Prilly/Lausanne, July 6–7, 1978
Editor: P. Baumann
1979. 54 figures. X, 240 pages
ISBN 3-211-81519-8    Wien-New York: Springer-Verlag

Springer-Verlag
Berlin
Heidelberg
New York